Like engend'ring like

Like engend'ring like

Heredity and animal breeding in early modern England

NICHOLAS RUSSELL

Bromley College of Technology

CAMBRIDGE UNIVERSITY PRESS

Cambridge

London   New York   New Rochelle
Melbourne   Sydney

CAMBRIDGE UNIVERSITY PRESS
Cambridge, New York, Melbourne, Madrid, Cape Town, Singapore, São Paulo

Cambridge University Press
The Edinburgh Building, Cambridge CB2 2RU, UK

Published in the United States of America by Cambridge University Press, New York

www.cambridge.org
Information on this title: www.cambridge.org/9780521306577

First published 1986
This digitally printed first paperback version 2006

A catalogue record for this publication is available from the British Library

Library of Congress Cataloguing in Publication data
Russell, Nicholas.
Like engend'ring like.
Bibliography: p.
Includes index.
1. Livestock – England – Breeding – History.
2. Livestock – Genetics – History. I. Title.
SF53.R87 1986   636.08′2′0941   85-21350

ISBN-13 978-0-521-30657-7 hardback
ISBN-10 0-521-30657-4 hardback

ISBN-13 978-0-521-03158-5 paperback
ISBN-10 0-521-03158-3 paperback

Contents

Charts

Tables

Plates

Preface

This book began life as a doctoral thesis[1] and what might be acceptable in that context is not necessarily adequate in work intended for publication. The doctoral candidate is a trainee scholar and what he produces is often a poor thing in literary terms. The examiners are forced to read his contribution, whether they find it stimulating or not. A wider audience rightly demands that such scholarship be both interesting and well presented or they will not spend valuable time reading it. Human nature being what it is and the task of recasting an entire thesis beyond my endurance, I am afraid that something more than the memory of its original form lingers on. I can only hope that the more turgid segments have been excised or sufficiently leavened to render them palatable.

My object has been to explore a biological example of the historical relations between science, technology and economic development. I have attempted little more than a simple narrative and cannot pretend that any profound explanation of events has been achieved or that this case history reveals any general insights into these relationships. Nor can I claim to have explored the subject of animal breeding in anything but the narrowest of contexts. I am only too well aware that intellectual and sociological questions are hardly touched on and that even the economic analysis might be considered perfunctory. These limitations reflect my own inadequacy for this task as a teacher of biology struggling to make sense of history, a discipline which calls for rather different credentials from those required in science.

Even at the level of a simple story what is written here can only be regarded as a provisional account of what may have occurred. To write a complex history of this type requires access to a mass of scattered and fragmentary sources, only a fraction of which I have had the time and resources to examine. My hope is that others, better qualified, may feel sufficiently interested to explore some of these sources and to correct or disprove the tale unfolded tentatively here.

I have made very full use of a number of standard histories of animal breeds and breeding and used the published findings of many biologists and archaeologists. I myself have made no biological contributions and nor have I anything much to add to the work that has been done on the detailed history of specific animal breeds. What originality this book may have lies entirely in the interpretation of existing historical data.

I have avoided using technical terms from modern genetics as far as possible but there are some places where their use is unavoidable. If their meaning is not clear from their context, most of them are defined in readily available reference works. I have not, therefore, thought it necessary to include a glossary of such terms.[2]

Finally my thanks to various librarians, archivists and their assistants who have been unfailingly helpful during my inquiries, to the members of the history of science and technology seminar groups at Imperial College between 1974 and 1979 for useful comments and criticism and to Professor A. Rupert Hall, Dr Marie Boaz Hall and Dr Norman Smith for their encouragement and support during what was an extremely long haul.

Introduction

Most historical analysis of domestic breeds of animal has concentrated either on their original domestication at the hands of early man or upon the development of livestock breeds in common use today, most of which cannot be traced back further than the early nineteenth century. Serious analysis of the events between these early and late periods has seldom been attempted. In the study of early domestication, questions about the effects that domestication may have had on the animals concerned and what processes of human domination or association were required to convert them from wild to domestic status are generally asked. For the later period most investigators have been breed specialists, asking such questions as: where did a given breed originate, how specialised and superior was it when compared with other breeds, and who were the great figures in its historical development? For the intervening period, either the questions are not relevant or the evidence does not exist that would allow us to answer them. However, questions of a rather different type may well be answered by the evidence that is available.

The whole period between early domestication and recent breed history is too complex to be dealt with within the pages of this book. Therefore, I have decided to consider the history of a restricted group of species, in England, between the sixteenth and the late eighteenth centuries. The species chosen are those with a central economic role in agriculture and transport: the horse, the cow and the sheep. The choice of the beginning of the period was governed by the earliest appearance of printed texts on contemporary technical and scientific matters and the survival of reasonable quantities of farm and estate accounts and other types of document relevant to this investigation, and that of the end by the publication of Arthur Young's journalistic portraits of the agricultural practice of England. This marked the opening of a period of comprehensive description of this subject by many writers such as William Marshall and the reporters working for the Board of Agriculture on the preparation of the County Reviews. This material has been so fully investigated already for its picture of agriculture between 1780 and 1820 that a re-examination appeared redundant.

What questions about the period under consideration does the available evidence allow one to ask? There are clearly two central queries:

(1) Did any significant changes in the economic performance of domestic species occur between the sixteenth and late eighteenth centuries? It will be argued that in some cases the extent of such improvements has been exaggerated but that, conversely, for some species there is evidence of changes in productive characters which may be interpreted as improvements.

(2) To what causes should one ascribe any observed improvements? The answers to this question are bound to be less satisfactory than those to the first because of the multiplicity of possible causes. Three broad possibilities suggest themselves:

   (a) Breeders deliberately modified the economic characters of their stock to make them more profitable or more attractive by the application of selective breeding procedures employed with the object of achieving such changes.

   (b) The changes were the consequence of actions on the part of breeders, in management or selection, which were not specifically designed to produce the results that occurred. They may have aimed at entirely different objectives and breed improvements may have been mere by-products.

   (c) The changes were the accidental side effects of improvements made in general agricultural technique during this period in areas other than those of livestock breeding. For instance, better animal nutrition may have led to the natural selection of breeds better able to make use of it.

None of these three possibilities may have been the decisive factor in any particular species at any specific time or place; all three may well have made some contribution. Only when one of the three was significantly more important than the others is it likely that its influence would be clear from contemporary records.

The actions of breeders under points (a) and (b) above could have been influenced by opinions about the theory of heredity, or, conversely, observations on inheritance amongst their animals could have led them to articulate principles to explain the transfer of characters from one generation to the next. Generally, however, it seems that explanations of hereditary and reproductive phenomena came from philosophers and natural historians and that breeders of livestock before the nineteenth century played little part in elaborating such theory.

The effects of domestication and the action of animal breeders on the form

Plate 1. Wild ancestors to domestic species: *Ovis musimon* (European moufflon). A ram of the flock kept at London Zoo in Regent's Park. This species is a native of Sardinia and Corsica. Either this species or one of its close relatives, the Asian moufflon of the near East or the Argali of central Asia, are believed to be the ancestors of domestic sheep. Author's photograph.

of domestic animals have often been discussed in accounts of the origin of domesticated forms. Zeuner was a leading exponent of one view that domestication arose from symbiotic associations between man and wild species of animal. Zeuner speculated about the successive steps in the process as follows:

(i) Loose contacts occurred between man and the potential domesticate. The breeding of the animal remained totally outside human control.

(ii) The animal was confined within the human environment. It bred in captivity but its breeding was largely unregulated by man.

(iii) Selective breeding was organised by man to obtain specific characteristics and occasional crossing with wild forms occurred, possibly to reintroduce desirable characters that had disappeared under primitive animal keeping.

(iv) Economic considerations led to the planned development of breeds with certain desirable properties.

(v) The wild ancestor was persecuted or exterminated once satisfactory domestic forms had developed.[1]

From point (iii) onwards Zeuner clearly envisaged that economics would take over from ecology and that the different varieties of domestic species were the results of deliberate decisions made by breeders. While not following the same stages in detail, Berry[2] also seems to make the same general distinction between an early, *animal-keeping* stage, where changes in the form of proto-domestic animals were the result of ecological and evolutionary processes that were not dependent upon human intervention, and an *animal-breeding* stage, where changes were a consequence of human decisions about the choice of desirable fancy or economic characters. In the animal-keeping stage Berry sees an arbitrary restriction of the total gene pool, since some alleles will be excluded from the proto-domestic sample. The total genetic variability of the domestic group will therefore be less than that of the wild population, but the conditions within the captive environment are likely to produce more phenotypic variety since the reduction in effective breeding-population size will increase the degree of homozygosity and show up recessive phenotypes that would be excluded under natural conditions. It is also likely that the captive group would be pre-selected for docility and perhaps small size, although the phenotypic effects of taming and poor nutrition could lead to the same results. Berry predicts only minor morphological change in this phase.

Once *animal breeding* with human intervention in mating began, he predicts a further fall in genetic variability as only certain character expressions would be thought desirable. He believes that for these desirable traits selection would be deliberate at this stage and predicts a progressive loss of heterozygosity as the allowed gene pool became increasingly restricted. If certain features of the animal were favoured morphological change might become marked as a result, not just with respect to the selected characters, but also through correlated changes in others as well.[3]

Thus both Zeuner and Berry see the effects of deliberate human selection as paramount in the development of diverse breeds which show variation in fancy and economic characters. Mason came to the same general conclusion, arriving at this position from a consideration of the physiological uselessness of some of the prominent morphological features of domestic breeds, so that he found it difficult to see how these characters could have been derived unless it was at the whim of human fanciers.[4]

An alternative view that modern forms of diverse breeds were not wholly the result of human decisions was first suggested by Darwin.[5] He was anxious to find in the artificial selection exerted by humans upon

Plate 2. Wild ancestors to domestic species: *Ovis musimon* (European moufflon). A ewe of the flock kept at London Zoo. Author's photograph.

domesticated animals a parallel series of very gradual changes over time to those which he believed had occurred in wild species under the influence of natural selection. He therefore postulated that exaggerated fancy points, such as the fantail character of some varieties of domestic pigeon, had not been derived from any deliberate human decision to select for such extreme character expression. The early breeders did not have in mind the final form to which the varieties would come after many generations of breeding. They had selected the best slight variants in tail character in each generation to breed from. 'Best' in this context probably implied those with the largest or most flamboyant tails. Darwin thought of the final form as being the result of continuous slight selection in one direction, the ultimate expression of which was the modern form. In this sense the human breeders were not consciously aiming at the extreme fantail form, but only at slightly exaggerated tail-fanning characteristics, so that the final form was an accidental consequence of selection policy. Thus Darwin would hold approximately position (b) considered above – that human intervention was significant in the breeding of domestic animals but that the types obtained over time were not necessarily the result of deliberate human attempts to create these types.

There are those who would go further than this and argue strongly that the modern diversity of domestic forms is entirely the result of persistent selective forces created by the mere existence of the domestic environment and its management constraints. They see these essentially accidental by-products of human management and breeding control as the main pressures operating on domestic breeds, to the exclusion of the consequences of deliberate selection for specific characters made by breeders. These fortuitous consequences of domestication may have remained the dominant controls over breed form long after the animal-breeding stage of domestication was reached.

For instance, Scott has speculated that the many breeds of dog which have developed since the original domestication of the species from a small wolf some 12,000 years ago, have resulted from an evolutionary adaptive radiation into niches created by the different cultural and management conditions that they have encountered during this period.[6]

His view of the original domestication resembles Zeuner's biological association, but he evidently believes that deliberate human intervention in the breeding process has had little to do with the subsequent specialisation of breeds. In some senses he seems to follow Darwin in the opinion that selection of the best examples in any generation for the specialist characters of a breed took place, but he also clearly sees breed development as an entirely 'natural' process in which deliberate human control over breeding has played no part at all. It is in this sense that alternative (c) above was suggested as a cause of change in agricultural species.

Some further confirmation of the effectiveness of the domestic environment itself as a force for breed change may perhaps be obtained from the one well-documented experimental domestication of a wild animal that has so far been attempted. The Norway rat has been subject to a redomestication begun by King in 1919 and discussed extensively by both Berry and Mason.[7] The Norway rat (*Rattus norvegicus*) was introduced into Britain as a wild animal in about 1730 and into North America in about 1775. It was probably taken into captivity to provide rats for baiting with terriers in about 1800. There appears to have been early selection for albino forms, which were used for breeding and exhibition. From these albinos have developed the modern strains of laboratory rat. No details of the process of domestication have been recorded.

The repeat domestication experiment used 16 male and 20 female wild-caught Norway rats, which became the 'captive grey' population. Problems were immediately encountered in that only 6 of the 20 females proved

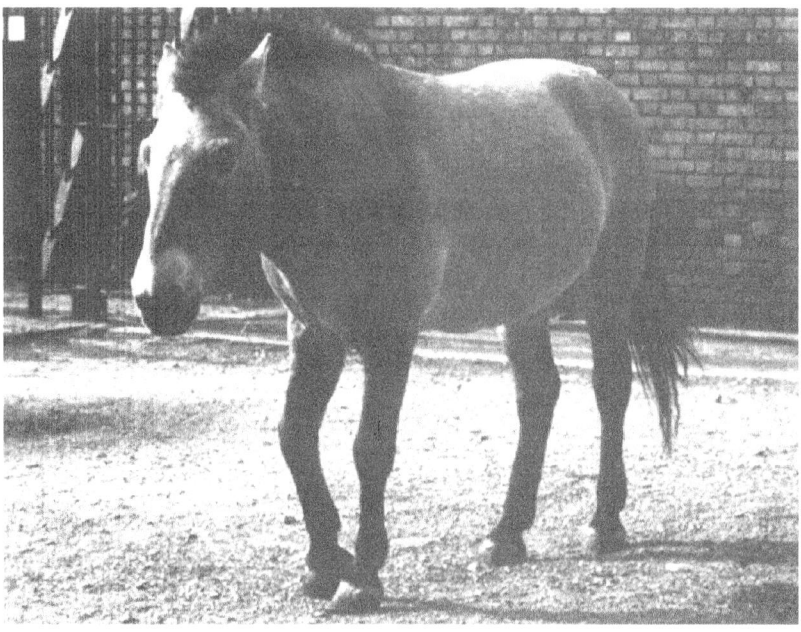

Plate 3. Wild ancestors to domestic species: *Equus przewalski* (Przewalski's horse). This specimen belongs to the herd at London Zoo. Przewalski's horse is the only surviving species of wild horse. Wild stocks were extant in Mongolia until at least the late 1940s. However, the species survives as a series of captive zoo populations, all derived from two base groups taken into captivity in Hamburg and Munich at the turn of the century. It is thought that either this species or the extinct Tarpan of the Ukraine are ancestral to the domestic horse. Author's photograph.

fertile. Even among those that would breed, offspring mortality was high and the first captive-born generation had to be fostered onto well-domesticated mothers to ensure survival. The infertility was attributed to fear and tension in the females, which led to the disruption of normal hormone patterns. Subsequently, female sterility declined to zero by the 13th generation. This experiment clearly could not mimic the events of the previous 'natural' domestication. It strengthens Zeuner's emphasis on the gradual biological association between man and the potential domesticate, rather than the direct exploitation by humans of captured animals. In the early associated state the animals would still breed freely while they would not in conditions of close artificial confinement.

Further changes in the domestic population of Norway rats followed in subsequent generations. The breeding policy adopted was one of selection of vigorous animals for breeding, a positive policy for a complex of physiological characters, which might parallel the primitive control exerted in the second phase of animal breeding, but which otherwise showed no deliberate selection for any specific traits.

The general population showed three changes over what, for large domestic animals, would be a long period of 25 generations. There was an increase in fecundity from 23 to 63 young per female (caused by the earlier onset of sexual maturity), an increase of 20% in body size and a decrease in aggressiveness. The body size change seems to parallel that which occurs in small, isolated subpopulations of wild species on islands, where a tendency has been observed for small species to increase in size and for large ones to decrease. There appear to be natural selective forces operating under these conditions, although how the changes in size of domestic animals might relate to this island phenomenon is unknown.[8]

The decrease in aggressiveness and the early sexual maturity both seem to be natural responses to the greatly changed environmental conditions of 'domestication', since no deliberate selection for these traits was adopted. Among domestic animals where human intervention is more marked, it has been pointed out that a potent selective force towards an early-maturing genotype will be set up by the tendency to breed from the earliest-maturing animals in any litter or group of offspring.[9] In this experiment, even where there was no indication that such selection occurred, the increase in fecundity as a result of earlier maturity was extremely marked.

Thus the mere change in environment and management produced by bringing an animal from its wild habitat to relative confinement and control seems of itself to produce strong selective effects on the population (or relax some natural selective forces) quite apart from any deliberate decisions made by human breeders.

Much more information about the changes that follow domestication may be obtained over the next few years after the experimental, and now commercial, domestication of other species of large grazing herbivores, especially the red deer in Scotland and New Zealand and the eland in South Africa. The change from hunting wild herds to ranched herd management has happened very quickly and one would anticipate a rapid development of domestic strains, especially as scientific breeding technology is being employed.[10]

The decisions that breeders have made in the past and continue to make now may be defined as breeding strategies. These are decisions such as whether to buy in or otherwise acquire herd or flock replacements from outside or to breed one's own replacement stock from the existing working group. If the decision to acquire from outside is taken, what policy of total or partial replacement should be employed and what breed or variety should be used? If the choice is made to breed from one's own stock, which animal

Plate 4. Wild ancestors to domestic species: the 'reconstructed' aurochs. The putative ancestor to domestic cattle, the aurochs, *Bos primigenius*, has been extinct since the seventeenth century. Descriptions and an illustration of the animals survive. Earlier this century the Heck brothers crossed several central European domestic cattle breeds together to produce a type which is said to resemble the aurochs in appearance and behaviour. This specimen, from the St Symphorian zoological park in southern Normandy, shows the characteristic dark body colour, white nose ring and long, straight horns. Author's photograph.

should be bred with which and why? Any given strategy or combination of strategies will not now, and did not in the past, necessarily produce the desired results because of the countervailing influences of poor management, poor understanding of heredity or the continued action of natural selection overriding the direction of artificial selection. It would seem to be sensible, before taking the discussion of animal breeding any further, to describe the range of strategies available together with their likely consequences. This will form the subject matter of the next chapter.

1

Breeding strategies

These strategies have changed little from classical to modern times and represent the restricted range of reproductive possibilities for breeding animals. Genetic theory has shown that some of these strategies are more effective than others under particular circumstances and it is now possible to discriminate usefully between methods suitable for different types and species of livestock. Decisions about which strategies to employ, and the effectiveness of any given strategy, have been vastly improved since the arrival of mature quantitative and population genetic theory, but the strategies themselves are still essentially the same. Scientific breeding programmes based on the application of this genetic theory are so complex and expensive that only large operators or national organisations can organise them effectively.[1] Where breeding is on a small scale, practice remains relatively untouched by modern genetics, for example in dog, cat and Thoroughbred horse breeding. Here the discrimination between strategies has not really improved since the eighteenth century and may, in fact, be worse as a consequence of the spread of the pedigree breeding system in the nineteenth century and the exaggerated respect that is still felt for this 'traditional' procedure by both breeders and customers alike.

The inheritance of complex physiological characters in animals is very difficult to analyse and is almost infinitely susceptible to environmental modifications which obscure the underlying hereditary pattern and led many breeders to suppose that inheritance played little part in the production of superior stock. This confusion between the inherited and the environmentally contributed worth of an animal has always made breeding a tricky procedure. There still remains considerable confusion about the genetic causes of physiological superiority, so that eighty years after the rediscovery of Mendel's work we cannot be sure that the theory upon which our current 'scientific' decisions on strategy are based is the 'correct' explanation. Under these circumstances, some of the healthy disrespect for genetic theory felt by many contemporary breeders may not be as entirely misplaced as it at first appears.

A *breeding strategy* may be rigorously defined as the degree and type of selection that a breeder of animals chooses to exert on the parents from which he produces his successive generations of stock. This supposes that there is a sufficient excess of animals over and above those necessarily marked out for economic use to produce a group from amongst whom selective breeding can occur. Poor management, disease, poverty or ignorance have frequently meant that numbers were too low for there to be any chance for the breeder to operate any selection at all. The consequences of a breeding strategy will be complicated by the effects of natural and accidentally generated artificial selection. These are produced by the degree of management exerted on the stock and the balance of adaptation between persistent aspects of the 'wild' environment and new aspects of the 'domestic' environment. These effects are often not susceptible to control, so that even a sound strategy of breeding can lead to consequences quite other than those intended. Deliberate objectives for which breeders might aim could be grouped under three headings.

(1) *The breeder might inadvertently produce a deterioration in the productivity, fertility and beauty of his stock.*

One must suppose that this can seldom have been a conscious objective of any breeder but it is nevertheless an all too frequent result of careless or misguided management. Among pastoral societies where livestock represents a currency, overgrazing and overstocking occur endemically, so that unproductive animals are an inevitable consequence of undernutrition. The introduction of animals productive in one environment or set of management conditions to another entirely different set is another recipe for deterioration. For example, the introduction of temperate, high-grade dairy cattle from northern Europe into the tropical climate and poor grazing of Africa was a disaster. On a less dramatic scale, the same kind of thing has occurred in all historical periods.

(2) *Since deterioration was such a common experience, one of the most frequently stated breeding objectives was to hold the status quo and prevent degeneration of stock.*

Once a subpopulation had been engineered, or more probably stumbled upon by accident, its maintenance became an obvious priority among the owners of this group. For most of the history of breeding this seems to have been the most ambitious objective aspired to. The pursuit of positive improvement over what already existed was rare because such a perception depended on a combination of historical understanding and a belief in positive change. Most believed that previous generations were better, grander and nobler than the present (for instance, the Renaissance perception of the superiority of all things classical) and most religions held the view that originally created things were perfect, so that any improvement was theoretically impossible. Thus both Christianity and Platonism stressed the originality and primacy of perfection and the inevitable degeneration over time that has occurred since the Creation. In the intellectual sciences this may have inhibited evolutionary thinking. To no less an extent, distorted versions of the same philosophies may have inhibited belief in livestock improvement. Adherence to this view would not logically preclude improvement back to the originally perfect form from the degenerate contemporary, but in practice this view was seldom held.

(3) *The breeder might believe the productive, economic, fertility or aesthetic qualities of his stock could be improved.*

This is the general objective of all European and American livestock breeding programmes today. We unconsciously tend to believe that such objectives were common in the past, although this attitude seems to have been largely absent before the eighteenth century. Most of the breeding strategies outlined below tend to be considered in the light of the positive improvements which they might produce, but one must bear in mind that, historically, arrest of degeneration almost certainly provided the dominant mental framework for breeders. Of course, any breeder might well employ more than one strategy at the same time.

Random mating

This implies that the breeder makes no effective selection of stock, so that the parents in any generation are a random sample of any flock or herd so far as the breeder is concerned. Of course, they may not be a random sample in the biological sense because such a casual breeding system often goes with minimal management, with the result that natural selection may operate strongly on the parent population, both in favouring the survival to adulthood of only some genotypes and in allowing full play to the behaviour patterns which are so important in the reproduction of large mammals. In practice, this complete non-interference in breeding must be very hard to achieve.

Negative breeding strategy

Here a selection of the worst animals is made to act as breedstock so that the best animals may be sold or used to realise the maximum profit. The worst are almost invariably selected by their physical appearance and condition and are adjudged poor by comparison with the objective of the breeder in producing beautiful or productive animals for sale or subsistence.[2] Such a strategy may have variable genetic consequences. Selection of the old as breedstock may not of itself be genetically detrimental, especially if the animals concerned have survived in the breeders' hands because they are economically productive. The diseased and the sick may also be employed, again with variable genetic effects, since if they are 'accidentally' sick no genetic poverty is implied. However, they may well be lame or sick because they are the weakest and therefore almost certainly genetically inadequate as well. If such a negative strategy is combined with poor management, which for obvious reasons it very often is, the result will be stock deterioration, lower productivity and lower fertility. In most cases where this is common it is probable that the breeders have little belief in inheritance as a determinant of animal form.

Such a negative strategy is buffered by natural selection. Whatever the strength of negative, artificial selection on the breedstock, natural selection will ensure the survival of the genotypes that can best cope with poor management. Natural selection will oppose the artificial selection exerted against productivity, fertility and health. The same natural selective buffering is frequently encountered when the opposite – positive selection for production – is aimed at. Natural selection may prevent improvement in productivity as much as it may oppose negative selection.

Positive breeding strategies

In these cases breedstock selection is based on some positive assessment of its worth. Animals will be selected which are either no worse than average or, in better managed systems, better than average. A number of procedures may be distinguished as follows:

Selection based on phenotype

Exclusion of the worst phenotypes
The simplest strategy is the exclusion from the herd or flock of the poorest offspring from the previous generation, or of the old, lame and sick stock. Here, culling of the stock by slaughter or sale is practised on all those animals not up to the standard of the average, so that even if the stock then bred from is randomly selected, the possibility of well below average stock participating in breed perpetuation is removed.

Selection of the better phenotypes
Only a restricted group is allowed to breed, as opposed to the whole group or a random sample from it, and this group is consciously selected because it is above herd, flock or breed average in the characters considered desirable, such as vigour, beauty, conformation, productivity and fertility. Frequently, however, one finds that suitability to local conditions and aesthetic or fancy

points are considered before productive characters. There seems to be an assumption that if these two areas are satisfactory, reasonable economic performance will follow. Economic characters, most of which are physiological, are often polygenic in their control, frequently in a manner more complex than simple additive action at several loci. The identification of the best variants in such productive characters is itself often difficult, calling for quite sophisticated measurement of various inputs to, and outputs from, individual animals. Even an apparently simple character, such as milk yield, cannot be satisfactorily assessed without complete annual milk records, preferably over several lactations, and this does not seem to have been achieved before the nineteenth century. It is therefore not surprising that selection for productive characters as a primary objective seems to have been so rarely practised. Where production was aimed at, some very simple morphological character, not always a good guide to the true economy of performance, was used. The most obvious example was size, either of the whole animal or of some productive part, such as the fleece of the sheep.

The implication of such selection must be that the offspring are expected to inherit the desirable parental characters. Until recently, selection by phenotype has often been misguided because the degree to which any character was strongly inherited, as opposed to largely determined by environment and management, has been impossible to determine. Such a distinction between an inherited component of character and a non-heritable, environmentally determined component would have been meaningless to many earlier breeders. As far as they were concerned, the parent animals passed on the characters that they exhibited at the time of copulation. All such characters were considered potentially heritable. Therefore many environmentally acquired characters were regarded as transferable, so that high condition was often favoured, for instance, because the condition was believed heritable. Similarly, age at copulation was often regarded as critical since the parents were thought to confer the weaknesses and conditions of their age upon their offspring. In many cases one sex, usually the male, was seen as critically important, so that parental selection was applied vigorously to one parent type only.

Provided the external appearance of an individual is a good reflection of its genetic constitution (the expression of the characters considered significant is strongly heritable), selection on the basis of appearance is not only sensible, but also genetically effective, in that better gene combinations are selected and the offspring of the matings will more strongly resemble their selected parental sample than does the general population. But useful selection may eventually cease as the best expressing genotypes become widespread in the population. Once this has occurred the correlation between offspring and parental quality may become poor as other factors, mainly environmental, become more significant in the expression of character.

But the breeder may not necessarily have envisaged a process of positive improvement and in many cases such improvement did not result from this strategy. He considered the selection in each generation as a check on the decline of his stock. Probably too few breeders ever selected consistently in one direction for long enough for the effects of such selection in improving the aesthetic or yield points to become obvious. Furthermore, as we have seen, the management system may not have allowed an improvement to take place since it operated its own independent constraints. At the present time, selection on the basis of the phenotypic expression of a productive character, such as milk yield or growth rate measured under standard conditions, is often used as a basis for selection in what is termed 'performance testing'.

Selection based on geography

Selection by place of origin
Since environmental effects were so often perceived as operating directly on the expression of character and such character modification was regarded as heritable, it was a logical conclusion to suppose that stock type was entirely determined by the climatic and nutritional features of the area where it lived and that if such animals were moved, they or their offspring would change permanently under the changed conditions that they experienced.[3] Therefore a breed was peculiar to a place and moving it converted it to the breed of the area to which it was moved. Longhorn cattle originated in Lancashire while Sussex had its own short-horned breed in the seventeenth century. A naive expression of this belief would suppose that if Longhorns were moved to Sussex they would, within a few generations, acquire short horns merely through the direct operation of what Darwin later called 'the changed conditions of life' on the reproduction and character transmission of the animals.[4] It is true that some superficial evidence to support this view could always be found since imported types of animal were seldom maintained pure but invariably interbred with local stock, so that their particular features were lost in those of the local population. Analogies were also drawn with plants in this context, where phenotype is notoriously variable in different soils and climates.[5] Holding this belief implied that if you wanted to maintain desirable features of a breed from a different region, constant import from that region was essential if the foreign stock was not to degenerate.

Selection based on hybridisation

Sire or group rotation

This is a system of reasonably constant import of sires of the same breed into any female breeding flock or herd from a different region or from a different subpopulation to prevent too long a continuation in a local herd without 'taking a cross'. This prejudice against continuing for long with closed population breeding, even if it was not markedly inbred or incestuous, seems to have been strong. As an alternative, parts of female flocks or herds were sold off and new stock acquired.

Creation of hybrid stock

This involves the selection of parent stock from different breeds, varieties or localities to try and combine certain desirable characters of each in the hybrid offspring. It can be effective only under certain limited conditions.

If the cross is remade with every generation from the basal parent stocks, the hybrid F_1 population is constantly re-created. This is both expensive and tedious and only characteristic of mature, stratified livestock industries, where specialist breeds from different regions are combined in intermediate environments and then perhaps hybridised again in yet other areas, the crosses being remade in each generation. An example is the modern British sheepmeat industry, whose stratification developed in the nineteenth century.[6]

As an alternative, the F_1 cross-breds may be persisted with and bred amongst themselves if very rigorous culling is practised. Since the hybrids will not breed true to type, all sorts of variants between the F_1 and the two parent forms will result and it will be necessary to cull off-types over several generations to produce a type approximating to the original hybrid that will be sufficiently homozygous to form the basis of a new 'pure' breed. This is also an expensive and wasteful procedure, not suited to the facilities of small breeders. It has been achieved successfully in the twentieth century in the creation of new sheep breeds in New Zealand and America, where the work was carried out on government experimental farms.[7]

Grading-up

This is a further variant on cross-breeding practice. The local breed is transformed into a foreign one by the repeated use of imported sires in the matings, so that with each generation the number of genes of the imported breed in the local population rises sharply until after four or five generations of consistent mating to the imported sires the local race is genetically 95% import type and only 5% local. This can only be a sensible policy if it is believed that imported stock will breed persistently to type in the new environment, and therefore conscious grading exercises are largely nine-teenth- and twentieth-century phenomena.

On a smaller scale, this technique may be practised in an attempt to 'grade up' to a dominant or successful sire by breeding to him for two, or sometimes three, generations. This may lead to the problem of close inbreeding with subsequent genetic depression. Grading-up on the breed scale usually involves the use of several unrelated sires to avoid this danger.

Selection based on breeding value

All positive selection systems so far considered have been based on the selection of phenotype. Common observation shows that such selection is often unsuccessful for a number of reasons, of which two of some significance are:

(i) The characters concerned may only be marginally determined by genes. Hence selection for the expression of the phenotype will be a very poor guide, since inheritance does not determine its expression to any extent.

(ii) The genotype aimed at and selected for may be selected against because of low viability in natural or domestic conditions. Deliberate and accidental selection may work against each other.

Therefore attempts have been made to find better guides to the nature of the offspring liable to be produced by any group of parents. Breeders have attempted to see behind the frequently deceptive expression of character in an animal and arrive at some independent judgement as to its real worth as a potential parent by attempting to assess its *breeding value*, the capacity to pass on to its offspring desirable economic or aesthetic characters which the parent itself may or may not express. Perhaps the most obvious example of such a potential is the spectrum of genes for milking capacity possessed by a bull. Milking genes are conferred by both bull and cow to their offspring, but of course the bull does not express any of these genes. Therefore systems have to be devised for testing the genetic potential, or breeding value, of a bull in a character which he cannot himself demonstrate. Two general procedures for selection by breeding value have been used extensively. These are *pedigree analysis* and *progeny testing*.

Pedigree analysis

The detailed parentage of an individual animal suggested for mating is analysed. In general it is argued that the breeding value of an individual is determined by its parentage over several generations, in that if successful, expensive or notable animals occur in its ancestry, it will produce better offspring than a similar individual whose pedigree is less distinguished because notable animals are missing from it. In extreme cases, even when the phenotype of an animal is clearly inferior to other potential breedstock, a distinguished ancestry will ensure its employment as a parent. The mature form of this system was fully evolved in the nineteenth century and has persisted to the present day. The origin of this generally erroneous belief in the value of pedigree is somewhat difficult to determine, but some of the following elements are undoubtedly involved.

Pedigrees had a practical value when the state of the livestock trade was confused. In the seventeenth century, complaints were rife that the horse stock of the country had become so mixed that the external appearance and markings of a horse were no guarantee of the qualities that should have been expected from a horse of a given type, and the horse dealers themselves were clearly not to be trusted in this context. In this kind of confused market, an established system of warranted pedigree would help the buyer to know what he was paying for. Partly as a consequence of this, an unofficial pedigree system did grow up in the eighteenth century among racehorses. But by the end of the century falsification had become so widespread in order that owners could enter horses for races for which they were not eligible by virtue of age or some other restriction, that James Wetherby was prompted to compile official pedigrees in the *General Stud Book* so that confusion and deceit might be avoided cheaply and easily. There was no hint that he saw the *Stud Book* as any sort of basis for breeding policy.[8]

There was also a firm belief in the persistence of nobility or family superiority in those descended from distinguished or successful historical figures. In all human societies where an aristocracy has evolved, the original members of the ruling class usually owed their position to some superior skill, power, good luck or criminal activity. Besides the creation of legal systems to bolster their entitlement to privilege in perpetuity, they have frequently attempted to justify their continuing privilege by arguing that the superior power and character of the ancestor or ancestors is hereditary, so that membership of the aristocratic family of itself entitles its members to privilege and power in their turn. The origins of this sociological phenomenon are beyond the scope of this discussion. The parallels between, on the one hand, the human obsession with title, hereditary position and social caste and, on the other, animal pedigrees, are too obvious to need emphasis. All pedigreed animal groups take their origin and claim to superiority from their descent from legendary ancestral beasts such as Bates's herd of Shorthorn cows or the famous trio of Arab/Turk horses in the pedigrees of all modern Thoroughbreds. The original selection of stock in the foundation herd may have been rational and based on the best contemporary performance, or the result of fashion, accident or the purest chance. Whatever the merits of the original group, the descendants supposedly continue to enjoy the character that made their ancestors famous.

This pedigree system has merit under two sets of conditions only. First, it may have some validity if no more than one or two generations are considered. Here, the presence of superior phenotypes can suggest superior breeding value among the present generation provided that the characters concerned are strongly heritable, so that the genotype of the individuals in the current generation may reflect some elements of the genotypes of their recent ancestors. This may have occurred at the origins of modern breeds, or rather, was perceived to be the reason for the presence of the foundation animals in the herdbooks. In effect this is a variant on phenotype selection where checks are made to see if close relatives in the previous generation had superior phenotypes as well. If further breeding is restricted to the offspring of a few foundation animals then purity of breed is ensured. Unfortunately this may not be related to genetic quality since only with a high degree of genetic homogeneity (which seldom occurs except after long inbreeding) are purity and genetic superiority likely to be linked in a given phenotype. Even when the foundation group is small, if no inbreeding has occurred to encourage homozygosity, the purebreds will be genetically heterogeneous and therefore of widely variable genetic worth. As so much resistance has always been shown to rigorous inbreeding on moral and observational grounds (since severe inbreeding depression often occurs even when associated with the rigorous culling of off-types) most pedigree herds of farm animals possess no genetic superiority by virtue of their ancestry at all, although both pedigree breeders and their customers believe that they do.

Secondly, a pedigree running over several generations can only be significant in highly inbred groups, where the genetic uniformity means that the genetic superiority of the foundation stock (if it possessed any) may persist in later generations. Very few breeds appear to be sufficiently inbred for this to be the case. The general belief in the pre-potency of the characters of distinguished ancestors has no basis, nor has the theory of reversion to, or reappearance of, ancestral forms following a cross of offspring of different families in the pedigree, which was much fancied as a justification for pedigree in the nineteenth century. Only in the Shorthorn and the modern dairy Friesian has inbreeding (not apparently accompanied by significant genetic depression), off-type culling and the superiority of the foundation stock been good enough to create a caste with a generally superior genotype. Here, breeding to pedigree has become important to ensure that the selected pattern is not disturbed or diluted by foreign or uncontrolled gene imports and, paradoxically, to ensure that too close inbreeding does not occur. Contrary to what many pedigree breeders imagine, in most closed breeds the very thing they require for continued improvement is some fresh gene input so that the new patterns can be selected for. Thus, except for a small group of successful inbred herds, closed herdbooks are a long-term disadvantage to most breeds since, instead of being modified in response to changing demand, old breeds merely disappear before newly created ones designed to meet a new demand. The successful inbred groups are rare, partly because inbreeding depression is so universal, and partly because it is now beginning to be grasped how much productive, physiological characters depend for their superior expression on heterogeneous gene combinations and not on classical purity and uniformity of inheritance of 'beneficial' genes as has been supposed until recently.

Progeny testing
Pedigree analysis has been shown to be a highly fallible way of judging the breeding value of potential parents in a contemporary generation. Theoretically, a far better guide should be provided by the progeny test. Here, breeding value is assessed by sample breeding of putative parents, whose offspring are then monitored for performance, under conditions of controlled management, to see whether parental phenotype superiority is reflected in the offspring. The test is usually only practicable for males, where a statistically valid sample of offspring may be easily produced. If the tested offspring are a good batch, the sire is used for widespread breeding, if not, then that sire is not permitted to breed. The testing is slow and expensive since a sample offspring generation have to be produced before real breeding can begin. It also shortens the reproductive lifetime of the individual so that such tests are only viable on a wide scale when artificial insemination is available. Furthermore, the difficulties of controlling the tests in terms of comparable offspring management and the contributions of the many separate females that have to be used, make these tests potentially deceptive unless the data are very carefully collected and processed. In many cases, testing does not justify its expense, so that many breeds are now planned more economically on one- or two-generation pedigree to assess their genetic worth, and on carefully recorded performance testing to obtain as accurate a picture as possible of the economic phenotype of the potential parent animals.

A final note should be added on the conditions under which breeding occurred, especially the time and place of copulation. The effects of heredity and environment were not really separated in the minds of seventeenth- and eighteenth-century animal breeders. The environmental conditions in which reproduction took place were regarded as just as important as the quality of the parent organisms. To the seventeenth-century breeder particularly, astrological conditions and the imaginative powers of the breeding organisms at the time of copulation seemed as significant as any other environmental constraint on the quality of the offspring.

It has been part of the Whig tradition of the history of science to see the survival of astrology and the alchemical pre-traditions of chemistry as mystical and irrational theories which the emergent sciences had to see through and reject. Astrology, magic and superstition were seen as essentially unscientific. More careful analysis has shown that it is historically incorrect to pluck out the newly emergent mechanical philosophy of the seventeenth century and separate it from the continuing organic and hermetic models of nature. The organic model conceived all natural phenomena as essentially biological and anthropomorphic, while the hermetic picture saw natural phenomena as controlled by secret, magical forces which could only be understood by those who could grasp the correct interpretation of the coded signs and symbols which these forces displayed in nature.[9]

The educated elite of the seventeenth century took astrology seriously. Planetary action was seen as a possible cause of many physical, biological and social phenomena.[10] At the practical level, the timing of reproduction to take advantage of favourable astrological conditions was carefully considered by some but regarded as nonsense by others. What little experiment was attempted proved equivocal, sometimes supporting the notion of significant planetary influence and sometimes not. It seems that in general the practical breeder who was concerned with the quality of his stock considered common sense, fallible guide though it often proved, superior to theoretical prediction based on philosophical analysis, in determining his breeding policy. He might not hesitate to ascribe his failures to witchcraft or an unfavourable alignment of the planets but he seldom sought success solely on the grounds of planetary influence and regarded the quality and condition of his breedstock as the most obvious determinant of offspring quality.[11]

2

◆◆◆

The classical tradition: theories of heredity and breeding practice in Greece and Rome

Writers of both theoretical and practical works during the European Renaissance seem to have been so influenced by the discovery of classical texts that almost the entire mental framework of the ancient Mediterranean appears to have been transferred to the rest of Europe between the fifteenth and seventeenth centuries. It is possible that ideas and techniques closely parallel with those of antiquity may have originated independently in early modern Europe. Even if the authors of books self-consciously sought classical authority as their guide, that is no guarantee that what they expounded was derived from that authority. However, conventional wisdom has it that transfer rather than independent discovery was the order of the day. Therefore no discussion of scientific and technical works in post-medieval Europe can be sensibly conducted unless this classical contribution is understood. A further point arises, of course, on the question of how much the actual practice of farmers or other technical men was based on book-learning. That is a theme which will be referred to often in subsequent chapters.

Throughout this chapter I have had to rely on English translations of classical works. Some of these are scholarly and reliable but others are not. This is a less than perfect procedure but the non classical scholar has no other reasonable alternative.

The vagaries of manuscript survival, translation and copying have led to the accumulation of many errors and omissions, so that it is doubtful if what has survived can be regarded as an accurate reflection of ancient opinion and practice. This does not matter greatly here since it is the seventeenth- and eighteenth-century perception of these affairs which is required.

Classical theories of heredity

It is something of a misnomer to speak of heredity here. We have seen that the modern distinction between heredity and environment as separate governing factors in the form of organisms did not apply in earlier times. It is also true that heredity was not differentiated from embryology in the way that it is today. Both the antique and the early modern naturalist aimed at an explanation of generation, the power of an organism or pair of organisms to produce offspring like themselves. The major phenomenon to be explained was the way in which offspring came to show this resemblance following the development of an embryo from reproductive contributions made by the parents. The mechanism and physiology of reproduction and foetal differentiation were the central points at issue. The detailed variations in the patterns of resemblance, which are the subject matter of modern heredity, were regarded as trivial and less exciting. There were exceptions. Aristotle was certainly concerned with the problems posed by these patterns, as were several seventeenth-century naturalists, but generally speaking the inheritance of character variation was given less attention than reproductive and developmental phenomena.

Theories of heredity from the earliest Greek philosophers have only survived as fragments or as reports or critiques made by later authors. These early works concentrated on the question of the origin of semen in the male, the presence or absence of a similar semen in the female and on the determination of sex. An association was made between the levels of 'perfection', or organisation, of an object or organism and its degree of innate heat. The more heat it possessed the more perfect it was. As in most patriarchal societies, maleness was thought to be more perfect than femaleness and therefore innate heat became associated with masculinity and strength. In reproduction, the volume and heat of the semen, or alternatively the degree of heat in the womb, were generally held to determine the sex of the offspring. Onto this was grafted another measure of perfection, the contrast between right-sided and left-sided, based on the observation that the majority of people of all races are right-handed, while the left-handed form a small minority of 'abnormalities'. Thus males became associated with right-sidedness and females with left-sidedness. When the human uterus was thought to be divided in two like that of other mammals, the right or left position of the embryo could be given as a cause of male or femaleness. Alternatively, it could be attributed to the product of semen from the right or left testicle and was increasingly so derived when the central position of the human uterus was later recognised. There were extensions of this theory, varying from advice as to which side to lie on after copulation, to the suggestion that the right or left testicle should be bound up if male or female offspring were desired and that, when animals were copulating, the side upon which the male descended would show what sex of offspring had been conceived.[1]

In the later surviving Greek texts, two general theories of inheritance were given: that found in the Hippocratic collection and that of Aristotle. Both of these were employed by later Roman authors, by the early Christian Fathers in their discussion of the transfer and entry of the soul, by medieval scholastic philosophers and as a basis for the revival of investigations into generation in the seventeenth century.

The doctrine of *pangenesis* seems to have originated in the speculations of the fifth-century philosophers, Leucippus and Democritus. A full statement of the Hippocratic version of this theory appeared in the twin treatises, *The Seed* and *The Nature of the Child*. Both male and female were held to contribute semen in the process of reproduction. The male contribution was obvious, while the presence of a female contribution of a similar nature was inferred from approximate parallels in the pleasure experienced by both sexes during copulation. The male semen was said to be derived from the most potent parts of all four fluids or humours in the body, the potency being demonstrated from the exhaustion produced by its release.

It was not clear if the abstraction from the four humours on their own was sufficient to provide all the necessary information for the next generation, and abstraction from the tissues and organs was invoked as well. The source of semen in the female was not explicitly stated. The pleasure of the female was said to be of a lower order, but more prolonged, because while the male secretion was produced suddenly and with violence, the female fluid was generated more slowly and less dramatically. The act of copulation warmed the fluid humours of the body such that a foam was produced from them. This was separated from the fluids in the brain. From there the spermatic fluid diffused into the whole body, but especially into the spinal marrow. The major route of transport from the brain to the spine was given as the veins running near the surface of the body behind the ear. From the spinal cord the semen entered the kidneys and in the male passed from there via the testicles to the penis. Specific details for the female were not given.[2]

The explanation of sex determination was rather vague. The basic cause of the sex of the offspring was the relative strength of contributed semens. If both partners gave a strong semen this would produce a male, if both provided weak semen it would generate a female. But when one partner produced weak and the other strong semen, the resultant sex would depend on the quantity of semen produced by each parent.[3] It was then proposed that since semen was drawn from all parts of the bodies of both parents, the child resembled both parents to some degree. The offspring could neither reproduce one nor the other parent precisely, but nor could it bear no resemblance at all. The similarity with respect to any one part was said to correspond to the quantity of semen from each parent for that part. In organs where the father contributed the greater amount, this would dominate the maternal contribution and that organ would resemble the male, and conversely for organs where maternal semen had the greater volume.

Deformity in the children of healthy parents was easily explained as a consequence of either disease in the womb, a womb that was restricted, or as a result of injury to the abdominal region during pregnancy. But for the production of sound offspring from deformed parents, the explanation was more difficult.

The children of deformed parents are usually sound. This is because although an animal may be deformed, it still has exactly the same components as what is sound. But when there is some disease involved, and the four innate species of fluid from which the seed is derived form sperm which is not complete, but deficient in the deformed part, it is not in my opinion anomalous that the child should be deformed similarly to its parent.[4]

The absence of deformity seemed to be explained because the semen was derived from the four humours, and not from the actual material arrangement of organs and tissues. Thus the diseased parent would merely have a one-humour deficiency in the semen, not in the arrangement of structural materials as they existed in the organs. And yet, elsewhere, resemblances to arranged parts, rather than mere humoral components, were used to explain the similarity between parents and offspring.

Embryo development was considered in *The Nature of the Child*. The semen from both parents mixed together into one mass through the effect of heat. The hot air in the centre of this mass forced a passage for itself through which it escaped. Breath from the mother could now penetrate the mass. As a consequence of this movement of maternal 'breath' inwards carrying nutriment, the seed inflated and formed a membrane round itself. Further growth was the result of nutrition from the mother supplied in the form of menstrual blood.[5] Thus there were no menstrual periods during pregnancy. When there was no pregnancy, excess fluid in the form of blood accumulated and flowed out once a month. A similar accumulation did not occur in the male because of his hotter and drier nature.

While the flesh of the embryo was formed from menstrual blood the differentiation into organs seemed to be controlled by maternal spirit or breath. The nature of the controlling process in this development and the mechanism whereby semen from the two parents gave form to the various organs of the offspring was obscure. The parts which had been completely mixed up in semen formation now apparently separated out by an attraction of like to like.[6]

Aristotle's theory of generation was both more complex and more intellectually rigorous.[7] Whereas the Hippocratics believed that both male and female contributed a similar spermatic fluid or semen, Aristotle held that only males produced this substance. The only obvious reproductive contribution from the female was menstrual blood. Aristotle argued that its presence excluded the existence of any semen resembling that generated by the male.[8]

Both reproductive secretions were described by Aristotle as 'residues'. Such residues were useful abstracts derived from the concoction of blood, being those materials which were in excess of those required for the nourishment and growth of organs. Semen and menstrual blood were merely two residues; other examples were milk, marrow, fat and external features such as nails, claws and hair.[9] Semen was seen as a residue of particular potency, both because it came from the more perfect male and therefore had greater innate heat, and also because the loss of a small volume during copulation was exhausting – an observation also made by the Hippocratic physicians. The semen was able to carry information about the form of the organism because it was drawn from the blood,[10] which carried nourishment to all organs:

and this, too, is why we should expect children to resemble their parents: because there is a resemblance between that which is distributed to the various parts of the body and that which is left over.[11]

Thus in contrast to pangenesis, where the semen carried an imprint from the organs, Aristotle's semen carried an abstract of what was going to nourish them. The nutrition necessary for the various parts of the body, in its concocted form, appeared to be a potential version of the organ before it arrived there to be converted into the actual substance of the organ. For Aristotle, this relationship between nutrient and nourished had to exist if the processes of nourishment and growth were to be explained. This potential, whether in the blood or in the semen, was seen as a spiritual not a material property, a *pneuma* or *dynamis*. Since the actual seminal fluid was not the important part, in Aristotle's view it was not even necessary for the material portion of semen to be passed over during copulation.[12]

The female residue was less thoroughly concocted and as a consequence was both greater in amount and more closely resembled the blood from which it was derived. Menstrual discharge was therefore the female reproductive residue.[13] Exactly how this residue differed from that of the male was never explicitly stated. Aristotle did not imply that the female menstrual blood was merely a formless matter from which the male *dynamis* would fashion the structure of the embryo.[14] Such a divorce of matter from form was a fundamental point of Plato's thinking, but Aristotle was never able to accept that formless matter or immaterial form could exist independently. As with semen, the menstrual blood contained a potential version of the organs abstracted from blood, but a potential of a lower, more material order than that of the seminal *dynamis*. If the female made no contribution in terms of form then all offspring would always resemble the male parent, but Aristotle was only too well aware that this was not the case.

The two contributions were judged to be the same in terms of carrying the lowest grade of nutritive and generative *pneuma* to produce the physical body, but the male contribution contained, in addition, a second-grade, sentient *pneuma* necessary for animal life and development. Aristotle regarded the character and behaviour of an organism as being of a different order from its mere bodily form. The male alone appeared to be responsible for the transmission of these features, while the female contributed the greater amount of information for the lower-order (in Aristotle's terms) provision of physical structure.[15]

In various places Aristotle likened the male *dynamis* to the actions either of a sculptor carving a statue from stone, or a carpenter making a bedstead from wood, while the female contribution was likened to the material from which the artefact was to be manufactured. This could be interpreted as the male providing 'form' or 'direction' while the female merely contributed matter. But Aristotle thought of the bedstead or the statue as being already 'potential' in the material, so that this could not be regarded as mere random substance. The artisan's task was only to release this form from its potential in the wood or stone. In modern eyes this seems to downgrade the craftsman and to give the material more credit for the final result than we are accustomed to. Read in the way intended, the analogy is very apt.

In the Hippocratic view, the sex of the offspring depended on the relative strengths and quantities of contributed semens. To Aristotle the essential difference between male and female lay in their relative ability to concoct semen. All the morphological and anatomical features of sexuality were seen as a consequence of this fundamental difference. The concocting ability depended entirely on innate heat, so that sex determination was controlled by the degree of this heat.[16]

Aristotle marshalled circumstantial evidence to support this contention. He claimed that young and very old parents produced predominantly female offspring because they were both deficient in heat. More fluid, feminine parents produced more female offspring. When the wind was in the south, greater fluidity was encouraged, and hence there were more female offspring. Therefore, to encourage more male children, Aristotle recommended copulation when the wind was in the north.[17]

The information transmitted during reproduction took the form of 'faculties' of different orders and types: faculties for sex, individual characters, the characters of ancestors, species characteristics and the features which produced an animal rather than a plant. Sex determination depended on whether the male seminal *dynamis* for the faculty of sex dominated the faculty for female development in the menstrual blood. If the female menstrual blood was well concocted for this faculty, and the male contribution for masculine development was weak, then the child would be female.

In the case of faculties for non-sexual characters, if the seminal *dynamis* for that faculty failed to dominate the maternally derived faculty, the result would be a 'departure from type' and a movement to the opposite, the faculty for the character inherited from the female parent. The clear implication of this was that the type or most perfect form of an organism would be a male, resembling its father in all its non-sexual characteristics. In so far as the semen failed to dominate the menstrual fluid there would be a departure from this ideal type, and any such departure was more or less imperfect or monstrous. In that sense, females, or the inheritance of any characters from a female parent, were degrees of monstrosity. However, they were clearly necessary and acceptable degrees, since offspring were very seldom perfect replicas of their fathers. Failures of any faculties in the seminal *dynamis* to master any specific female faculty would thus lead to the observed mosaic of resemblances of a child to both its parents.

Thus one form of interaction between the reproductive contributions led to a 'failure' of the semen and a departure from perfection towards the character of the female parent. Similarities between both parents and their offspring were explained. The rational basis of the explanation was very different from pangenesis, but nonetheless bore a close resemblance to it at the mixing or combining stage of the contributions. But how were similarities to more remote ancestors to be explained? Aristotle suggested that in some cases where a faculty in the male semen did dominate the female faculty, it might not manage this without being damaged. He said the consequence of this 'thing which cuts getting blunted by the thing which is cut'[18] was to cause a relapse from the faculty of the direct parent to the next one in importance, that of the next previous ancestor, that parent's own direct parents. Thus an offspring might not resemble its father in some faculty where there had been minimal seminal mastery, but its grandfather or great-grandfather, or even some more remote paternal ancestor. Similarly, if the faculty that dominated was female, it too might be damaged and would relapse, so that resemblance was to the grandmother or some remoter maternal ancestor. Frequent relapses would cause the offspring to show only minimal similarity to their parents and, if the relapses were of different degrees, so that the resemblances were mixed through several generations, the recognisable relationships would be just about nil.[19]

Aristotle believed in epigenetic embryo development, a gradual taking-on of form by organs as they grew, explained teleologically as the embryo being pulled along a pathway towards the mature organism. In some places he suggested alternate heating and cooling and the action of a connate *pneuma* as the instrumental forces involved, but he implied that the bulk of the nourishment would be supplied by the menstrual blood, giving it a dual role as seminal and nutritive substance.[20]

These two Hellenic theories of heredity were not modified to any extent by subsequent Greek or Roman writers although some practical medical and biological investigations of great importance did occur in Hellenistic Alexandria with the work of Erasistratus and Herophilus in the third century BC and the anatomical work of Galen in the second century AD. But neither these nor other authors produced any revision of the theoretical basis of generation.[21]

Animal breeding in ancient Greece and Rome

The earliest references to animal breeding in Greece appeared in Xenophon's work, *On Hunting*. He advised resting brood hunting bitches before they were mated and suggested that the resulting puppies should be nursed by their natural mother and not by a foster bitch since, 'nursing by a foster mother does not promote growth, whereas the mother's milk and breath do them good and they like her caresses'.[22] Xenophon gave little advice on parental selection, merely advocating the use of a 'good' dog. Comments on sporting animals also appeared in the contemporary *Republic* of Plato. Plato's object was to use hound breeding as a model for eugenic policy. He noted that even well-bred dogs seemed to show variable quality. He believed a sportsman would only breed from those animals judged the best, not from the inferior, since qualities were generally thought heritable. The discussion was simple and the 'best' in this context were taken as those in their prime, not the youngest or oldest. Although this prejudice against youth and age was common in antiquity, the context suggests that the selection of breedstock was not made on the grounds of age alone, but that there was a selection of those animals which were good hunters.[23]

As a basis for his discussion, Plato almost certainly had in mind the Spartan constitution drawn up by Lycurgus, recorded by Xenophon as the *Constitution of the Lacedaemonians*, which contained the following eugenic principles. To produce fine children, it was suggested that women should not be sedentary but at the peak of physical fitness to ensure that when married to strong men the resulting children would also be tough. Men should only be allowed to take wives in the prime of their lives so that the weaknesses of youth and old age were not transferred to their children. Intercourse was to be infrequent so that the physical passion expressed would again encourage the production of vigorous children. Xenophon's suggestion that mother bitches were better nurses to their whelps may have derived from the same source, since Lycurgus believed free-born women should devote themselves to motherhood to the exclusion of all else to ensure their children received as much attention as possible.[24]

These measures were not based on any procedure which would produce genetic selection because the designer of the system, despot though he might be, had not envisaged control over the choice of breeding partners for his subjects. The eugenic mechanism depended on maternal care and the most rigorous attention to the physical and mental condition of the parents, reflecting the *a priori* belief that the offspring would inherit all the structural and physiological features of the parents at about the time of intercourse.

Aristotle gave a considerable amount of detailed information about domestic animals scattered through the nine books of the *Historia Animalium*. It seems unlikely that he had any personal experience of agriculture, although he seemed quite familiar with cattle-ranching in Epirus and, as a dweller at aristocratic courts, he had firsthand knowledge of hunting-dogs and horses.

He believed that obesity was a cause of sterility and, since older animals were more likely to be overweight, this was a contributory reason for their poor breeding performance, together with the general fall in vitality that came with increasing age.[25] His reasoning was that both blood and body substance were derived from food and that if the body was fat, there would be little material over for the manufacture of blood. Since semen was derived in turn from excess blood, the absence of any excess restricted the amount that could be produced.

For the majority of species, apart from the general theme that generation was feeble in old and young animals, Aristotle gave no specific age optima for breeding in females. Nor was it forthcoming for rams or dogs, but for the males of other domestic species definite primes were given as to when the offspring were thought to be of the highest quality. For boars this was very young, not beyond three years, but in the stallion the quality of the sire improved up to twenty years of age, and in the bull, up to five years. This opinion was not based on any observation of the quality of the offspring but was inferred from the degree of interest that the males showed in the females.[26] The extreme size and lethargy of fully mature bulls and boars, for instance, was taken as a sign of the flabby and lethargic offspring they would generate.

Aristotle's information was most comprehensive for horses. The breeding behaviour of both mares and stallions was said to be markedly different from that of most other species in that both sexes were generally more ready to breed. In-oestrus mares were said to be more salacious than sows or cows and Aristotle ascribed the Homeric legend of mares being impregnated by the wind partly to this cause. The general prejudice against old males for service was missing in the horse. Mares came to maturity at five years old and stallions at six, but Aristotle was fully aware that both sexes would breed before this and he even advised that sexual maturity could be speeded up by high-plane nutrition, a technique commonly employed for agricultural species at later dates.

In Hellenic times, horses were used exclusively for sporting and military purposes by the Greek upper classes and were therefore subject to more attention than any agricultural species. There seem to have been two distinct management systems employed, one for 'ordinary' cavalry horses raised in troops and a second, stable-based system for high-bred, aristocratic mounts.

The troop breeding system was based on groups of twenty or thirty free-ranging mares. This stud was kept away from their stallion except at breeding time, when he was introduced and allowed to serve them at will. The same stallion was apparently used on the troop for season after season and some degree of inbreeding was bound to occur as a consequence, as the breeding life of a stallion could be from fifteen to twenty years.[27] This attitude seems odd, given the general prejudice against close inbreeding reported by Aristotle himself in various anecdotes about camels and mares.[28]

In the alternative stable system, mares were only bred from once every two years and Aristotle went so far as to suggest that an interval of four or five years between foals would be even better. Clearly, such a slow method was only appropriate for the most valuable animals. He suggested that foals bred in this way would be of better quality than those bred from a mare every year. A rational basis may have been that the foals were run at foot for several seasons, benefiting from the mare's milk, or anthropomorphic parallels with the Spartan constitution may have been in his mind.

These matings were individually controlled in a stall or yard and Aristotle spoke of a valuable 40-year-old stallion still used for service, even though he had to be lifted up for the operation. The system was practised in private stables and the mating could be organised outside the usual breeding season provided that the mare was restrained. It is unlikely that incestuous breeding of the type encouraged in the troops was common in aristocratic stables. Aristotle gave no hint that any parental selection occurred, but one might speculate that the eugenic preference for the correct time and conditions of mating was more important than phenotypic or performance selection among variants in the population.[29]

Aristotle also discussed cattle-breeding at some length. The rich pasture lands of Epirus, on the western Greek mainland opposite Corfu, were famous for the large size of their domestic animals.[30] The cows were supposedly so large that milkers had to stand to reach their teats. This sounds somewhat unlikely until the emphasis placed on large, ceremonial bulls in classical

Greece is remembered.[31] The bull was the chief Greek magic beast, and Zeus himself had the alarming habit of parading about in bovine form. Aristotle constantly emphasised milk production when talking about cattle and it must be assumed that northern Greece and perhaps Macedonia were specialised pastoral regions.[32] The cattle of Epirus ran in loose herds, and were allowed to follow their natural preferences on the ranges, running in single-sex groups except in the breeding season:

The bull in breeding time begins to graze with the cows, and fights with the other bulls (having hitherto grazed with them), which is termed by graziers 'herd-spurning'. Often in Epirus a bull disappears for three months together. In a general way one may state that of male animals either none or few herd with their respective females before breeding time; but they keep separate after reaching maturity, and the two sexes feed apart.[33]

During the season cows usually became unmanageable. The general pattern sounds only a short step from nomadic pastoralism, and the control over the breeding and selection of stock must have been minimal. Their fierceness and the seasonality of their breeding both suggest that these animals were under only minimal supervision. But there must have been cattle under closer management. Aristotle gave some hint of this when he suggested that cows were impregnated in a single mount, but that if fertilisation had not been successful 20 days should be allowed before resubmission.[34] This implied a degree of control over mating impossible under Epiran ranching conditions.

In contrast to the sparse Greek literature, there are a reasonable number of surviving works on agriculture from the Roman period. Opinions differ as to the reliability of these sources as evidence for Roman practice contemporary with their composition. Many of them acknowledge a large number of earlier Greek and Roman sources, and even those which do not are clearly based on the same material. The later authors repeat wholesale sections from the earlier texts.

The first Roman writer to discuss animal husbandry in any detail was Varro. The manuscripts of his three books on agriculture are in poor condition, with many corruptions, so that caution must be exercised in using him as a source.[35] Although he claimed to be the owner of large stocks of sheep and horses, his account of breeding practice was based on conversations with Greek cattle owners in Epirus.[36] Evidently in the first century BC, Epirus was still a famous cattle-ranching area and what Varro described may be taken as an account of Greek rather than Roman practice and compared directly with Aristotle's earlier description.

Varro was clearly impressed with Greek cattle and the Epiran system of management. Cows were kept short of food for a month before mating while the bulls were kept away from the females and fed well for two months prior to their service.[37] He recommended the same principles for rams and ewes, claiming thin females would conceive more readily than fat, an idea which was common in Greece as we have seen.[38] The males were fortified because copulation in the male was supposed to be so exhausting.

Two bulls were kept for every sixty or seventy breeding cows, of which one was a yearling and the other a two-year-old. The cows were at least two years old, and preferably three, before they were allowed to breed. They were said to breed satisfactorily up to the age of ten years or even longer. The emphasis on bulls up to five years old found in Aristotle has now changed to a pattern of extreme youth in bulls. An exact parallel occurred in seventeenth- and eighteenth-century England, and perhaps for the same reasons. Varro had emphasised the desirability of getting all the cows into calf in the shortest possible time so that they would all fall 'at the most temperate season'.[39] This could only be achieved in open-ranging herds with very young animals whose sexual vigour was very high. Aristotle would have discouraged the use of such animals on the grounds that they would generate weak offspring, but breeders probably favoured the younger bull because they believed his vigorous sexual appetite would be transferred into generally vigorous offspring and because they would gain the advantage of bunched calf drop.

Varro listed a number of conformational and 'fancy' points for an ideal cow and bull and showed a distinct preference for black cattle over red, dun or white animals. These last he regarded as very delicate, while black was supposedly the hardiest. This prejudice was to be repeated constantly in Europe in the early modern period. Both parents in any mating should be carefully selected for their conformation and should be from a good breed, as Varro clearly thought that quality was heritable.[40]

Aristotle's account had the cattle of Epirus roaming free with the bulls fighting each other. Even if selection was exerted on the females by culling out the least desirable, the opportunity for selection among the bulls would have been minimal. Yet Varro's informants were now anxious to show the importance of the bull in breeding, and this may represent a major change since the fourth century. Another, practical, reason for emphasising young bulls, now that they were being attended more closely, was that Epiran cattle were notably large, so there may well have been problems in handling mature bulls. Care of the cows at calving time also seems to have been improved as they were stall-fed and run on home pastures before and after calving. It was even recommended that the stalls had stone floors to prevent foot rot. For the remainder of the year they were best pastured on woodland where they could feed on undergrowth and foliage.[41]

Varro's account of cattle breeding, if it may be trusted, shows a reasonably careful control of mating, an ideal type of animal posited, and animals as nearly resembling this ideal as possible used for mating. The ideal was couched in purely conformational and 'fancy' terms, and the omission of any udder or milking points probably meant that dairy and mothering qualities were not deliberately selected for. This is surprising as one of the points singled out for the goat was that it should have 'rather large udders, so that it may give a greater quantity of milk and a richer quality in proportion',[42] and one of Varro's suggestions for breeding bitches was that they should have even, well-formed teats.[43]

Progressive breeding is evident in some of Varro's discussion of other species. As a general principle he stated that if too many young were born, some should be weaned to allow better growth of the remainder on maternal milk.[44] He gave as specific examples the removal of half a litter of pigs[45] and the weeding-out of the worst pups from a litter so that the best ones gained from the better food supply.[46] In this second case, selection of the best juvenile phenotypes was being practised, combined with favourable nutrition to enhance their advantage.

Varro commented at some length on the choice of foundation stocks of sheep, goats, swine and asses. They were to be selected from regions noted for the particular species concerned. A flock should be bought from one source and not in dribs and drabs from separate places, since the animals would be more contented, always having run together. Varro gave no opinion on the permanence of varieties once they had been removed from their original habitat, but hints elsewhere in his work suggest that he favoured heredity over environment as a primary determinant of breed type. A series of statements indicate that Varro's contemporaries were quite sophisticated in their understanding of artificial selection.[47] In swine the desirable character of prolificity was to be sought out and tested for in the choice of sows to retain for piglet breeding. There were hints of progeny testing for both pigs and sheep. In the latter species the stock being bred were to be monitored for the conformation quality of their offspring as well as the phenotype of the parents. These were not progeny tests under controlled conditions but merely general suggestions that the real value of any stock for breeding purposes could not be gauged until its ability to generate a population of offspring of similar quality had been recorded.

There cannot be much doubt that the general principle of looking at breeding value as well as phenotype was clearly implied by Varro. The recognition of the importance of breeding value suggests that he had a firm belief in the hereditary control of many characters in domestic animals. For instance, one should select sires in sheep and goats that had been generated by dams who were consistent twin-getters. Varro believed that the inheritance of this trait could pass through the male as well as the female line, so that a sire could passively carry this character of twin-generation, which he did not himself express, and pass it to the females with whom he was mated so that they would, in turn, produce twins.

Some further features of Varro's attitude emerge from his account of horse breeding. It now appeared to be standard practice to bring the animals together in a yard or stable, a procedure which had not been universal in Greece, although breeding remained seasonal.[48] Considerable emphasis was placed, following Xenophon, on judging colts, in an attempt to predict from the juvenile what the adult would be like. The conformation of the colt was considered and some attempt was made to judge its character and courage as well.[49]

The second large agricultural work surviving from the Roman period is the *De Re Rustica* in twelve books by Columella, composed in the middle of the first century AD some hundred years after Varro. Columella lived most of his life in Italy and owned four farms, three in Latium and one in Etruria. One quarter of the text was devoted to viticulture, while the second most important topic covered was animal husbandry, which occupied about one tenth of the work.[50] Careful reading of the text suggests that he had direct experience of some species, but he also relied heavily on earlier sources including, of course, Varro.

Columella provided the clearest statements since Aristotle of the reasons for rejecting aged or very young animals as parents, and more sophisticated reasons for 'steaming up' males before breeding to reinforce their physiological features to a fine pitch at the time of reproduction.[51] Varro had said that certain regions produced better breeds of livestock. Columella also believed this to be the case and followed the Hippocratic authors and Aristotle in ascribing most of the differences between varieties of animal to the environment and climate of the regions in which they lived.[52] Thus he recommended that introduced breeds of sheep should be brought from similar environments to make them suitable for local conditions, although he thought it even better to stick with the native breed of any region.[53] While he was an environmentalist on the question of the forms adopted by local breeds, he nevertheless believed that potential parents should be chosen carefully. He allowed some weight to heredity and believed the male to have the dominant hereditary role.[54]

Columella discussed the problem of inheritance in stock breeding more explicitly than Varro by dealing with the selection of suitable animals for the breeding of mules.[55] In this special case he saw that phenotype (appearance) and breeding value (ability to generate valuable offspring) were distinct properties which were not necessarily related, and warned all would-be breeders of the uncertainties of choosing a jack by its appearance. It was not just in the matter of form that the jack sometimes failed to shape his offspring like himself. Often the inheritance of colour seemed poor. Sometimes the key to colour inheritance could be seen in the hairs on the eyelids or ears, where the colour of these minor features became the body colour of the offspring. He argued here by analogy with the established opinion that a ram, although white-skinned, if in possession of a coloured tongue, would beget colour-skinned lambs.[56] On occasion even the parental extremities did not seem to give the correct colour, so Columella fell back on the suggestion, 'that the colour of the grandsire is transmitted to the second generation mixed with the elements which form the seed'.[57]

Columella used inheritance patterns from grandparental sires elsewhere. For instance he believed that the onegar (wild ass) could be usefully employed in the generation of mules, but not in the first generation. Wild onegars should be crossed with tame asses to obtain a hybrid form of jack for crossing with domestic mares to obtain mules in the second generation. Any first-generation cross of an onegar to a mare was thought to generate offspring that were too fierce and lean. The mule produced by the recommended second generation cross was said to inherit its useful qualities from the original wild sire and the derived half-tame jack.[58]

Columella stressed the particular dominance of males in colour-marking their progeny in his famous account of the experiments of his uncle in Spain, who used wild rams of an imported North African type with coarse fleeces but magnificent colour patterns. They were crossed, once they had been tamed, with ewes of the jacketed, fine-fleeced Tarentine breed, which produced hybrid rams with wild colour markings and also a persistent, wild-type coarse fleece. When these were crossed in turn with Tarentine ewes, they were said to have produced young rams with the wild colour pattern but the fine fleece of the domestic type and that the offspring of these rams in their turn bred true for the fine fleece and the wild colour. Columella made no mention of selection or culling among the offspring, which would have been necessary to ensure a reasonable connection between these two desirable properties.[59] It is significant that the grandsire and sire inheritance patterns reported here were identical with those described in the use of wild onegars for mule breeding.

This review of antique breeding practice and hereditary theory demonstrates that there was confusion over the relative contributions of heredity, physiological condition and environment in the control of domestic animal quality and that there was little useful interaction between the art of the breeder and explanations of the process of generation. The breeder could gain very little help from theory since the latter had no useful predictive value. Where attempts were made to explain practical phenomena, as for instance in Aristotle's theoretical explanation for the supposed sterility of aged animals because of their obesity, the connections were made at a relatively trivial level.

Neither pangenesis nor Aristotle's generative theory could guide the practical breeder. Columella's consideration of three-generation character transmission in asses and sheep would not have benefited from Aristotle's theoretical discussion of how characters could pass to filial generations from grandparental or other ancestral individuals. On the strictly technical level of breeding policy, it seems that Roman authors may have realised the importance of hereditary transfer as a significant determinant of animal quality and had noticed the rather different point that hereditary virtues were not necessarily obvious from external appearance. They were starting to search for methods of assessing what is nowadays referred to as breeding value.

3

Generation and the market: the background to animal breeding in the seventeenth and eighteenth centuries

Having considered basic breeding strategies and classical views on generation and animal breeding, I shall summarise the external factors which might have influenced farmers and breeders, before going on to discuss the details of breeding policy in England in the early modern period.

The animal breeder used his skills and techniques either to satisfy a market for products from which he could make a living or to compete for the possession of animals outstanding for their beauty or performance. The market was both for the animals themselves for sale to graziers, fanciers, dealers or sportsmen and also for animal products, with farmers both raising beasts and exploiting their consumer potential for milk, meat, wool and so on. The breeder's raw materials were the stock he could acquire by inheritance, purchase or home breeding and the insights and skills that he brought to choosing, mating and managing his stock. He came by these technical skills from observation, formal instruction, casual conversation and reading printed books on the theory and practice of agriculture. These last literary sources became available in English on an increasing scale in the sixteenth and seventeenth centuries and it is from them that it is possible to gather something of contemporary opinion on agrarian technique. In this book the emphasis will be placed on the selection and mating of stock and I shall only deal in passing with animal management, although obviously the latter had an effect on the results of breeding policy.

In a sense the animal breeder could have represented a bridge between the economic realities of the market place and the theoretical principles of reproductive and hereditary biology. In fact biological knowledge did not seem to underlie his practice to any extent. Nonetheless, it would be useful to see how the ancient heritage of ideas on generation was adapted and modified in Tudor, Stuart and Hanoverian times and to ask why it contributed so little to stock breeding. On the other side of the coin, although the breeder might not know any natural philosophy, he was directly regulated by the shape and size of the markets for his products. Therefore some overview of the agrarian economy of this period is necessary. It will be the task of this chapter to review these two matters briefly.

Theories of generation

In the extensive medical and philosophical literature of the fifteenth and sixteenth centuries, the Hippocratic origin of semen from all over the body was married to Aristotle's thesis that reproductive substances were built from excess nutrition, the refined particles of which were equivalent to the organs and tissues that they were going to nourish. Each sex produced a distinct semen of some sort, although there was no general agreement as to whether both semens were equally important. During the seventeenth century anatomical investigation was carried far enough to show that these antique theories of generation were untenable.

As we have seen, the term 'generation' embraced what are today two separate disciplines: heredity and embryology.[1] In modern terms these are different modes of expression of the same basic material, the gene complement. This has both to be expressed in the cells and organs of any living system so that differentiation and function may be controlled, and also to be packaged and distributed to the next and subsequent generations so that these will be organised on the same basis. It is only in the last two decades that it has been possible to understand how the hereditary units, the genes, can carry out both functions, using essentially the same explanatory theory. From the seventeenth century onwards, it seems that at any one time only one aspect of generation could be satisfactorily explained by current theory and that the other had largely to be ignored or explained in entirely different terms. For example, in the twentieth century, between about 1910 and 1960, the growing comprehension of the concept of the gene and its function was concentrated almost entirely on its hereditary properties. Understanding of embryological development in the light of this hereditary theory was virtually non-existent. Conversely, it was probably true that from the late seventeenth to the late nineteenth centuries, generation was understood in terms of the mechanics of reproduction and embryo development, and that heredity was largely ignored. It is only since 1960 that a single set of explanatory principles has been able to bring the two disparate strands of generation together again after centuries of divorce.

Since the hereditary aspects of generation suffered a theoretical eclipse during the period considered here, practical breeders looking to natural philosophy for guidance could find very little relevant to their needs. Their observations in turn had very little influence on natural philosophy. Only in the early seventeenth century had the two aspects of generation not yet drawn apart and philosophical speculation at that time contained much argument drawn from heredity. By the late seventeenth century, heredity was all but ignored as the thrust of experimental investigation in the maturing discipline of comparative anatomy brought forward extensive evidence about embryological development.

The scale of possible interaction between the learned and practical professions had been strictly limited before the seventeenth century by the lack of vernacular texts on theory from which the 'uneducated' man could learn the rudiments of classical and modern scholarship. During the seventeenth century there was an explosion in the publication of books in English with large print runs and a wide readership, although editions of even the most popular works were small by twentieth-century standards. In the past this has led many to question the relevance of either practical or theoretical books in the dissemination of knowledge, and to retain the view that the great majority of people in the seventeenth century only learned through verbal contact. Recent work has begun to show that this picture is erroneous and that books in both theoretical and practical subjects played a very considerable role in the life and education of all those who could read.[2] For example, James Master, a Kentish gentleman and occasional horse breeder, recorded in his diary that he purchased a copy of Sir Thomas Browne's *Vulgar Errors* as soon as it was published in 1646, a cheap manual of *Human Anatomy* and, most significantly for the discussion here, a *Discourse Touching Generation* for 1s 6d in 1664.[3] At that date this text may have contained some discussion of heredity, but a similar commentary published towards the end of the century would have been full of the new triumphs of comparative anatomy in the discovery of the events of embryo differentiation. These investigations were to lead in the early eighteenth century to the widespread adoption of theories of generation that depended on the concept of *emboîtement* or pre-existence. Such theories had no relevance to the animal breeder dealing with problems of variation from mating to mating among his domestic stock.[4]

The discipline of comparative anatomy had been developed to a fine art at Padua late in the sixteenth century and the most significant revelation from the application of its methodology in the early seventeenth century was the discovery of the circulation of the blood, made sometime around 1620, by William Harvey. By applying the same methods to the study of embryology, Harvey also took the critical steps in the growth of the embryological emphasis in the explanation of generation when he published *Exercitationes de Generatione Animalium* in 1651, made available to a wider audience through its translation as *Anatomical Exercitations Concerning the Generation of Living Creatures* in 1653.

Harvey's major contribution was to show the basic error of all doctrines of generation derived from antiquity by the repeated demonstration that neither masculine semen nor menstrual blood could be found in the uterus of fertilised females in the mixture known as the 'conception'. If this amorphous mixture did not exist, then neither Hippocratic nor Aristotelian generative theories could be accurate explanations of what occurred.[5] Furthermore, Harvey could find no trace of any initial material contribution by the female in the uterus, nor any sign that the male semen ever actually found its way there. Since the embryo of the viviparous mammals clearly had to originate somewhere, the female was the more obvious of the two potential material causes. Therefore he adopted the idea that the female contribution in the mammal was a primordium of similar type to that found in the early development of oviparous animals, such as the hen, and designated this primordium an ovum, although he made it clear that he had never been able to see an actual ovum and nor did he believe the mammalian ovaries had any role to play in their production.[6]

This primordium or ovum was triggered into development by the action of some spiritual male principle derived from the semen, not the semen itself, since Harvey denied that it could penetrate the uterus of either birds or mammals.[7] The embryo then developed from the ovum by a process of epigenesis, the gradual appearance of the organs and parts in an ordered sequence. Many of his contemporaries preferred the proposed alternative pattern of development, that of precipitation. Here, the organs in the primordium were present in miniature from the very moment of conception and embryo development was merely seen as the growth of parts to their full size. This growth without differentiation had the advantage that it did not require any differentiating force to be at work while the embryo was growing. A weakness of Harvey's theory was that he could not provide any satisfactory explanation of how epigenic development was controlled. He believed that nutrition was converted to blood and that from the blood the other organs were differentiated by the effects of its vital heat under the influence of a designing principle. The only such principle that he could envisage was God, so that he was forced to suggest that God's action was necessary for the development of each and every embryo for all species.[8]

Harvey believed that his work had rendered all older theory on generation obsolete and yet the book did not create any great stir such as *De Motu Cordis* had done 23 years earlier, nor did it seem to have any great influence on the anatomists who followed.[9] It seems that contemporaries saw Harvey's views as a mere restatement of those of Aristotle, since the emphasis on blood, vital heat, the heart as the initial embryo organ and the non-material contribution of the male all appeared to derive from Aristotle, whom Harvey went out of his way to present as his master and guide. His rejection of the menstrual blood as the female generative substance and his frequent criticisms of Aristotle for internal contradictions may have seemed less prominent at the time.[10]

Despite these criticisms of the master, fellow anatomists were right to see Harvey as a vitalist Aristotelian, a position which was already out of date before *De Generatione Animalium* appeared. The newer, mechanical philosophy was built on a model of nature as a machine and not an organism. God's role was to design and create a perfect system which would then run automatically according to the divine plan. Contemporaries such as Kenelm Digby criticised Harvey's vitalist need for constant divine intervention at every conception on these grounds.[11] The microscopical anatomists who followed, such as De Graaf, Swammerdam and Malpighi, all worked within this mechanical paradigm. Harvey's attitudes must have looked impossibly old-fashioned to them.

In the 1660s and 1670s microscopical investigations revealed the presence of follicles in the mammalian ovary, which were assumed to contain ova, and an accurate description of the early development of the hen's egg was published. In both cases it seemed clear that the early embryo arose from the female reproductive contribution only. Once again no obvious role could be seen for male semen. Rather than speculate about vitalist directing forces, the anatomists suggested that the ovum was the sole origin of the embryo and that the male contribution was merely some sort of trigger to set development in motion.[12]

Conversely, Leeuwenhoek and others had revealed the presence of spermatozoa in the seminal fluids of humans and other animals. The first published report was made by Leeuwenhoek in 1677. Many supposed these animalculae might be instrumental in reproduction, although the possibility that they were parasites or mere stirring units remained standard explanations for their presence among those who would not accept their primary reproductive role. The origin of the embryo from a seminal animalcule was seen as a logical alternative to the ovist position.

Thus, by the 1680s, alternative ovist and animalculist theories had been proposed and attempts to see both sexes contributing materially to the origin and differentiation of the embryo were abandoned. Instead, each alternative hypothesis was carried to its logical conclusion, not only to the exclusion of any contribution from the opposite sex but to the rejection of the parent organism itself from any direct responsibility for its own offspring. These preformationist and pre-existent theories of generation would have been untenable if any serious reference had been made to the evidence from biparental heredity. But such evidence was simply not considered, as the anatomists, by the nature of their investigations, were concerned only with the arguments from embryology and conceived generation solely in terms of individual expansion or differentiation.

The concepts of preformation and pre-existence derived from the controversy over whether embryo development was epigenetic or precipitate. The problems of epigenesis have already been rehearsed while the converse precipitation (or metamorphosis as some would have it) meant that the anatomists had nothing to 'explain' in embryo development apart from growth. The anatomists had some evidence for organ precipitation from the work of Malpighi on the hen's egg and Swammerdam on metamorphosis in insects. However, some contemporaries and many later writers went further and suggested that this was evidence for preformation, that all organs or organ precursors already existed, preformed within the parent organism.[13]

Some modern authors claim that preformation in no sense logically followed from precipitation, pointing out that precipitation was applied to the post-fertilisation of the embryo while preformation referred to the pre-fertilisation stage of the ovum or animalcule.[14] But the original ovists could see no fertilising function for the male semen, so that the post-fertilisation primordium was in no material sense changed from its pre-fertilisation state as far as they were concerned. Preformation could then logically follow from the acceptance of embryo precipitation. Once preformation had been conceded the logical extension, mechanistically and theologically sound from both Cartesian and Christian standpoints, was that the preformed embryo had pre-existed for many generations, ultimately that all the animals in the world now and to come had been pre-existent as tiny forms in the earliest animals created. There had only ever been one Act of Creation and God had directed the development of all animals from the very beginning. The constant interventions of God required by Harvey had been replaced by a single, bold, initial act, so that all future generations were packaged up within the first members of each species, a position which became known as *emboîtement.*

Many modern scholars find the parallel between the mechanical model of the unwinding clock of the physical universe and the unwinding clock of biological development irresistible as the intellectual framework of the age willing the mature *emboîtement* theory onto investigators.[15] The jump from De Graaf and Malpighi to the first version of the *emboîtement* thesis was made by Malebranche in 1674, but the step from precipitation to preformation may have taken some time to be generally accepted, so that adherence to pre-existence on a wide scale was an eighteenth-century phenomenon, perhaps stimulated by the growth of the alternative animalculist theory.[16]

Emboîtement should not have withstood criticism from biparental heredi-tary evidence, but very little serious attempt was made to marshal such material. Animalculists were better able to cope with it because they could argue that the female contribution from early embryo nutrition could modify the form of male semen. Ovists were on occasion forced to reintroduce the concept of maternal imagination in order to explain away the occurrence of some of the more monstrous deviations from the expected pattern of development. But there were more positive reasons for the failure to use arguments from heredity than mere lack of knowledge on the part of anatomists. The significance of variation was minimised in an attempt to find fixed criteria for classification.[17] Platonic philosophy lay behind much early work on classification and behind the search for the divine plan, with variation from the abstract 'type' regarded as mere accidental monstrosity, a view strongly reinforced by the acceptance of either a theory of *emboîtement* or the popular feeling that the biological clock was running down and creating imperfection and off-types.[18] In this sense seventeenth- and eighteenth-century taxonomists, as well as anatomists, held views which ill served practical workers among breeds and varieties of domestic plants and animals. The variations involved were liable to be dismissed as monstrous, corrupted by man and no part of God's design. The huge variation and evidence of complex inheritance of the variants through crosses in domestic stock could be ignored as irrelevant.[19]

The seventeenth- and eighteenth-century English breeder was therefore left in a vacuum by the biological theory of his day. From the 1680s to the end of the period under discussion here, preformation, *emboîtement* and fixed species classification systems dominated theoretical biology. The only authors to discuss hereditary problems and devise cross-breeding experiments with domestic plants and animals to try and illuminate the hereditary process were French and their work generally inaccessible to breeders and farmers in England. The only English writers to discuss the subject were those who wrote on generation in the seventeenth century before preformation came to dominate the scene. Three such authors were Sir Thomas Browne, Sir Kenelm Digby and Nathaniel Highmore. Towards the end of the century John Ray stood out against the application of mechanistic principles to biology because, as a field naturalist, he was only too aware of the problems posed by variation and inheritance for pre-existent theories of generation.

Sir Thomas Browne's popular *Pseudoxia Epidemica* or *Vulgar Errors*, published in 1646, contained some discussion of aspects of generation in its somewhat random refutation of various classical and folklore stories and opinions. The most extensive discussion of generation was on the origin and persistence of the colour of negroes. This most obvious variation and its inheritance, together with other discontinuous and frequently pathological human variations, were the main pieces of hereditary evidence brought forward by opponents of pre-existence in the eighteenth century. Browne accepted the view that men had not been originally created black. Some natural phenomenon had originally turned them black and this character had become heritable. He could find no reasonable cause for it and was forced to consider as an analogy the tendency for some wild and domestic species of plants to vary spontaneously, without apparent cause.

Browne then rehearsed the various theories of negro origin which had been proposed and, although unable to draw any sensible conclusion, he was absolutely clear that the negro character was now heritable and current environmental conditions had nothing to do with its expression nor any direct influence on the seed.[20] The case of the negro and the European provided one of the few well-known and thoroughly observed cases of characters which looked as if they should be a direct result of the environment, but which observation showed were not modifiable except through hybridisation between races of different character expression. It was thus destined to be discussed interminably in the eighteenth and nineteenth centuries, with few able to accept that it was an entirely heritable phenomenon. In less well understood cases of variation in nature and domestic breeds it is not perhaps surprising that no conclusions could be drawn either and that many, including Darwin, would argue well into the nineteenth century that environmental change could alter the form of an animal or plant by changing the nature of the seed which it transmitted.

In the same year that saw Browne's book published, a much fuller treatment of generation was provided by Sir Kenelm Digby as part of two large treatises on the physical and spiritual universes.[21] Digby's opinions on generation originated from the classical dispute between pangenesis and Aristotle. Digby was an Aristotelian believing in epigenesis and essentially holding Aristotle's view on the origin and function of semen. But he rejected Aristotle's suggestion that epigenesis was powered by a vital force or Harvey's opinion that it was divine guidance. Digby believed that development was induced mechanically by physical forces operating on the growing foetus. He had earlier expounded the view that such physical forces alone were responsible for the precise formation of consistent inorganic shapes such as salt crystals and he extended the argument to cover the differentiation of biological systems. His espousal of such mechanical forces and rejection of classical pangenesis left him struggling to explain some well-authenticated examples of inherited mutilations.[22]

In 1651, the same year in which Harvey had published *De Generatione Animalium*, the physician, Nathaniel Highmore, composed a treatise whose essential purpose was to refute the opinions on generation proposed by Digby.[23] Highmore believed in the preformed embryo before the fusion of seed. He was aware of Harvey's work and since its conclusions disagreed with his own preformationist position he found a number of reasons for rejecting Harvey's ideas.

Highmore began his refutation of Digby by attacking his weakest point, the control of epigenic development by the action of simple physical forces. How could the uniform forces of nature, such as cold, heat, wind, calm, gravity and so on, produce the diverse forms of the multitude of different species of plants and animals? Highmore reverted to the view that the form of the embryo must be determined by the parents in some way. The seed produced had to contain some information about the structure of the parent organisms but not in the traditional pangenetic way, although he gave examples of inheritance in domestic animals which were compatible with pangenetic mechanism.

Myself also have seen a kind of poultry without rumps: which breeding with their own kinds, still brought forth chickens wanting that part: If with others, sometimes they had rumps, sometimes but part of a rump. And not long since I saw a mungril Bitch, that had her tail cut close to her body almost, whose whelps were half without tails and half with tails.[24]

He did not regard the latter example as particularly strong evidence for pangenesis because the same bitch in the following year 'brought forth all with long tails, as she had before the cutting off'. In addition, there were far more cases of the mutilated producing perfectly sound offspring than the other way round since 'Spaignels, whose tails are always cut, bring forth whelps whose tails need as much cutting as their dams or sires did.'[25] Yet some version of pangenesis would have been the best theory of those extant in the seventeenth and eighteenth centuries to incorporate hereditary phenomena. Both Buffon and Maupertuis in the eighteenth century developed variants on pangenesis to integrate embryological and hereditary phenomena[26] and its espousal by Darwin in the nineteenth century speaks for its explanatory potential. But it was repeatedly rejected, even in its preformationist form, in favour of pre-existent *emboîtement*.

Highmore did not find Digby's use of Aristotle's superfluous nutrition theory of the origin of semen any more convincing than pangenesis. He reverted to a physiological theory of nutrition close to Anaxagoras in which nutrient substances consisted of a great variety of atoms from which those relevant for growth were extracted by digestion into the blood, which circulated them to the organs and tissues, where the atoms attached themselves to the biological material already present. Atoms were also stripped off the organs and tissues and left the body through breathing and excretion, so that the materials of the body were constantly turned over. Highmore suggested a theory of testicular (and here he meant both male testes and female ovaries) abstraction of spiritual atoms from the nutritional selection made by the blood. These more spiritual and 'forming' atoms were then deposited in solid matter to prevent their dispersal and loss of vigour, the whole concocted and stored up as the male and female seed. Highmore's position on nutrition compelled him to take a preformationist view of embryo origin because, unless the organ types of the adult were already there in the seed, before fertilisation, the correct nutritional particles could not be attracted to the correct organs either in the seed or the developing embryo. But Highmore did not believe in unisexual pre-existence but held firmly that seed from both sexes was essential. He followed Aristotle on the nature of the contributions with the male side composed of more 'spiritual' atoms for organisation, while the female atoms were more 'material'. His theory left open the problem of the inheritance of specific characters, such as rumplessness in chickens, since seminal atoms were not drawn from the actual organs themselves but from the appropriate nutrient atoms before they came into contact with their target tissues. He was forced, as was Digby before him, to conceive of the blood communicating an impression of the organs to the seed when a specific, variant character was inherited. Quite apart from his opposition to Digby, Highmore also produced arguments to counter Harvey. Turning evidence from heredity to his advantage, he was able to dispose of several notions held by Harvey, such as his supposition that females could respond via imagination to the visual presence of males.[27]

The great naturalist, John Ray, was particularly anxious to find a satisfactory theory of generation to fit with his understanding of biological variation. He was acutely aware of the amount of superficial variation within natural and domestic species of plants and animals and, unlike many of his contemporaries and the rational naturalists of the eighteenth century, he did not regard this variation as merely trivial and could not therefore simply ignore it.[28] In addition, as a theologian, any theory of generation which he adopted would have to take account of the properties of the souls appropriate to man and other types of organism and conform generally with Christian doctrine. He therefore had the greatest difficulty extending the mechanical, clockwork model of nature to biology, especially to the question of generation. He dismissed Descartes's attempts to bring mechanical philosophy into this field because the latter's theory depended on the notion that the male seed was physically united with the egg, with which idea Ray could not agree. Ray accepted completely Harvey's conclusion that the male contribution was spiritual rather than material, and followed him too on the divine nature of the force directing epigenesis. He did not find the action of uniform forces such as vital heat, or Digby's external effects, sufficient to explain the diversity of animal and plant form or embryo differentiation.[29]

For some biological followers of Descartes, such as Gassendi or Willis, the soul of animals other than man was often defined as material and their biology as mechanical. Only man had a rational soul for which some direct divine intervention was necessary. But Ray found the idea that higher animals were devoid of rationality absurd, given for instance the behaviour of the domestic dog. Descartes's own solution to this problem was to extend the presence of the same rational soul from man to the higher animals as well, a concept which Ray did not find theologically acceptable.[30] Even in the development of plants Ray would not concede control wholly by mechanical forces, since the movement of a plant from one soil to another did not, to his mind, alter its type or species but merely the details of its proportion. Something was directing the epigenetic development of the seed into the mature plant, some heritable self-knowledge or some divine force made it of the same basic type as its parents and the mere mechanical action of the substances in the soil or the effects of the climate could not be the only factors determining the form of the mature plant.[31] Ray found divine intervention and vital forces necessary at every turn to explain biological phenomena.

It was not only the universal presence of biological variation which puzzled him but also its inheritance. He could not ignore the arguments from heredity as could the comparative anatomists. He found this a particular problem in the light of his acceptance of Harvey's belief that the male seed was only a contagion and yet appeared to have a dominant role in the formation of offspring since the mule, for instance, resembled its male parent, the ass, more than its dam, the mare.[32] Consideration of the mule led him to the problem of the barriers that existed between species and the limitations thus set on the inheritance of variation. Why were species hybrids so rare and why was it that in the majority of cases where they did occur, they were sterile and could not carry forward the inheritance of variation? Why was it not possible for these interspecific variants to be inherited in the same way as the variations within species? Like his vitalist predecessor, Harvey, Ray 'came to a stop' over the question of generation. For Harvey, the problem he could not satisfactorily explain was the vitalist system of controlling embryo development, while for Ray it was the failure of mechanistic explanations of generation to take account of either the huge amount of biological variation or the known facts of its inheritance.

Harvey's own theory of generation, however, Ray could not accept despite his great admiration for its author. He conceded that much of the criticism of Harvey by mechanical philosophers was fully justified. Reluctantly, he was forced to accept mechanistic ovism and *emboîtement* as the only reasonable hypotheses, although he refused to raise these to a status in his own mind above that of working conjectures. He probably found some comfort in the theological rectitude of *emboîtement*, which may have compensated for its philosophical defects in his mind.[33]

Ray had no compunction in using the biology of domestic organisms as a model for the behaviour of natural species. He did not adopt the view that such forms were degenerate and therefore atypical of 'real' nature. He pointed out that the large number of forms of many domesticated plants were not species but only varieties since the forms would freely interbreed. By analogy he suspected that the description of external appearance alone was a poor guide to the species status of natural variants as these too could easily be mere varieties and not true species. He therefore believed in the adoption of biological criteria in taxonomy and thought that the presence of breeding barriers should delineate the limits of species. This would make the construction of any taxonomy difficult, since variant forms would need to be checked for their biology in the field, as well as having their morphological form described. Not surprisingly, this basis for classification was not adopted and still has not been fully accepted, despite the soundness of the biological arguments in its favour.

On the question of the origin of variation, Ray again used examples from domestic animals and took the conventional position that the origin was to be found in the environmental conditions in which the organisms lived. For instance, Ray believed the white colour of Arctic animals was caused by their constant inspection of snow and that the hairy pasterns of Flemish draught horses were a consequence of some environmental condition to which English horses were not subject, so that the latter had clean pasterns.[34] He also followed those exponents of animal breeding who thought breed types were entirely determined by environment. For Ray no variety could be called constant. He did not, like Thomas Browne, feel doubtful about this species variability in different environments because of the constancy of negro and European characters. Nor had he the growing experience of animal breeders that in many cases breed characters were not plastic but strongly innate and heritable, even after more than one generation of transfer to a new set of conditions.[35]

Ray's overall position on generation would appear to have been as follows. Species possessed a group of fundamental characters whose general form had been laid down at Creation and which had subsequently been unfolded via ovist *emboîtement*. The detail of these characters was widely variable in expression through the different environments in which organisms of different generations found themselves. Thus varieties within species were transmutable, although the species 'type' was fixed from its origin. He did not see environment operating in a pangenetic manner on the form of the seed, but working directly on the organism during its epigenic development. In that sense he would probably have disagreed with those animal and plant breeders who believed that environment was more important than heredity in controlling the form of an animal or plant and adhered to a pangenetic explanation of such changes in form as occurred when organisms were moved to different habitats.

Ray's confusion over inheritance may be taken as symptomatic of the problems that both naturalists and breeders from the late seventeenth century onwards had in applying mechanical theories of generation to their practice, were they sufficiently interested and educated to give the matter any thought. In turn, the disparate and piecemeal experience gained of how characters were inherited or not, represented too scattered and diffuse a source of data from which mechanical views of generation could be attacked.

The market for animals and animal products

Agricultural production certainly increased in Great Britain between 1500 and the end of the eighteenth century. How great this increase was, its precise timing, and the nature of the technical and economic forces which allowed it to occur, remain matters of controversy. It could have been achieved in a variety of ways. The amount of land managed for agriculture might have been augmented, existing techniques might have been applied more intensively, or perhaps husbandry was improved by developments in tools, livestock, crops or management systems. Any combination of these might have been important at different times. An increase in managed land area may have been significant in the sixteenth century, but it is clear that in the seventeenth and eighteenth centuries there must have been technical changes to achieve the improvement in output.

Farming is an economic activity and the great majority of landlords and tenants have always had to make a living of some sort from their land. It is a truism of economics that people will attempt to make the best use they can of their resources to maximise their standard of living. It is often claimed that they will have most incentive to innovate, invest and adopt new procedures under conditions of rising economic expectation and opportunity. These are said to be provided by an expanding demand for goods in a buoyant market. Conversely, static, depressed or shrinking markets are said to produce retrenchment and stifle initiative.[36] This may be valid for the adoption and spread of inventions (innovation) but seems doubtfully true of the novelties themselves. Hard evidence that market forces somehow call technical developments into existence automatically is not great. Inventiveness in science and technology may be independent of any direct economic causation. However, in agriculture in this period it is probably only the widespread adoption of novel techniques that one can hope to trace, for the occurrence and causation of the original inventions themselves are likely to have gone unrecorded. If it is true that such innovation requires market growth, then the spread of technical improvements in animal husbandry must have followed from a growth in demand for animal products. What were these market forces and how powerful were they in the three centuries under consideration here?

The state of both agriculture and its markets is hardest to understand in the early period from 1500 to the mid-seventeenth century because the sources necessary to illuminate them are scarce. The interpretation of agrarian events in Tudor and early Stuart times therefore remains an area of heated controversy where the outsider intrudes at his peril. The most prominent economic phenomenon of the period was the inflation in both prices and wages which began in about 1520 and did not finally stabilise until the 1650s. By twentieth-century standards this inflation was slow and relatively insignificant but, measured against the price stability that seems to have preceded it in the fifteenth century and followed it in the eighteenth, it was remarkable and requires explanation. It certainly worried contemporary politicians and pamphleteers, who cast about widely for both explanation and cure.

The inflation in industrial prices was much less than for agricultural products. Between 1520 and 1650 industrial product prices rose by a factor of three while between the same dates agricultural price inflation was approximately seven-fold.[37] The comparative picture for industrial wages is not so clear, but it appears that they rose less fast, perhaps keeping pace with industrial prices but falling well behind the agrarian price rise.[38] Two general explanations have been offered for this inflation. The first, which was fashionable until the 1950s, ascribed it entirely to monetary causes, the debasement of the coinage by the Tudors in the first half of the sixteenth century and the extensive import of South American bullion into Europe thereafter. However, it became increasingly clear that on their own monetary mechanisms could not be made to tie in closely enough with the known chronology of inflation.[39] The second explanation suggests that the basic cause was an increase in population. More people created a greater demand for food in the first instance and it was agricultural prices which experienced the main rise. According to this theory the agricultural sector was unable to respond fast enough to demand, and scarcity increased food prices.[40] The agricultural sector also supplied many raw materials for industry and its infrastructure, many of which were used in primary industries such as clothing and footwear. As the population increased, demand for these products must also have risen and yet the inflationary problem was not so great.

This explanation of increasing food prices as a consequence of population pressure outstripping agricultural production shows strong parallels with the orthodox view of the agrarian crisis of the fourteenth century.[41] Ultimately, this crisis was solved by the Black Death and the sharp fall in population that followed in the latter half of the century. In the sixteenth and seventeenth centuries the problem was apparently solved by agrarian and industrial innovation, so that the population was fed and provided for from home-grown resources, and, in the eighteenth and nineteenth centuries, by increasing external trade and the exploitation of resources, eventually on a world-wide scale. This second general thesis suffers from two difficulties. First, it must be shown that the chronology of population increase was compatible with the chronology of inflation and second, if this proves to be the case, that there was a causal connection between them and that the association was not merely accidental.

The latest evidence on population increase seems to support the association between inflation and population rather closely. J.C. Russell proposed some time ago that the English population on the eve of the Black Death in 1348 was about 3.75 million and it then declined as a consequence of the Death and subsequent plagues to 2.1 million by 1400. By 1430 population recovery was well under way and Russell estimated that by 1545 the population had recovered to 3.22 million. By comparison, an estimate for the French population around 1500 was 15 million and the Italian population in 1550, 8.85 million.[42] This confirms the sparse population in early Tudor England recorded by foreigners visiting the country. The latest estimates from Wrigley and Schofield suggest a rather smaller English population in the 1540s, 2.8 million in 1541, from which figure the population rose to some 5.3 million in 1656, after which it stabilised until 1731 before starting its phase of exponential growth to carry the population to 21.5 million by 1871. If Russell's medieval data are regarded as satisfactory, the picture of population growth is one of slow advance from about 2 to 2.8 million between 1400 and 1540, which represents about a 40% increase in 140 years, followed by a dramatic 85% increase over the next 120 years, after which there was a period of stability for a further 80 years. While the 40% growth from 1400 to 1540 was against a background of severe underpopulation and sparse resources, the more rapid growth from 1540 to 1650 was associated rather closely with the Tudor and Stuart inflation, especially if foodstuffs are taken as the marker, since the chronology of agrarian and industrial price rises was different and agrarian inflation did not really take off until the 1540s.[43]

The first difficulty of the population model of inflation would seem to have been overcome, although it is clear that the vicious Tudor debasement of the coinage in the 1540s did contribute to its origin. The second still remains. Was the inflation caused by population growth and, in the case of the agrarian sector, what were the economic and technical problems faced and overcome by farmers? On this question there are broadly speaking optimists and pessimists. The optimists suggest that agriculture did respond sufficiently to the increased demands to allow the population to be fed and the industrial base to increase from home-grown resources between 1540 and 1650. There were problems, symbolised by inflation and political agitation and the confused clash of social classes in a community moving towards a capitalist economy, but on balance the conditions of Tudor and early Stuart life showed no deterioration compared with the fifteenth century.[44] The pessimists, on the other hand, see the inflation of agricultural prices as a disaster for the poor, whose real living standards dropped sharply, with their position further weakened by political oppression from enclosing landlords and an exploitative new capitalist gentry. For the majority, this century was one of economic depression and severe political unrest culminating in the upheaval of the Civil War. So far as the pessimists are concerned the conditions of life were better in the fifteenth century and the supposed improvements of the sixteenth century were largely a mirage caused by the widespread use of the printing press giving more easily accessible documentation about the sixteenth than the fifteenth century.[45]

Sources available for the period from 1650 to the end of the eighteenth century are more revealing of agricultural innovation, but the economic mechanisms which might have powered such change are, if anything, more obscure than for the early period. There was no one dominant process like the great inflation upon which to focus. Before 1650 there is evidence of considerable economic incentive for food producers but only equivocal information about how they responded to it, while after 1650 the general outlines of agrarian innovation are clear but the economic motivation seems absent. The pull from an increasing population was missing and did not reassert itself until the 1740s, while wage levels only rose slowly. In the last quarter of the eighteenth century both population and wage levels started to accelerate upwards and the production changes associated with the Industrial Revolution began, although these increases alone are never considered a sufficient motive to generate the industrial transformation.[46] More agrarian and industrial produce was exported in the late seventeenth and eighteenth centuries, but the overall size of this export trade was tiny compared with the home market, so that its direct influence on economic growth, although more significant than its size alone would suggest, cannot have been the predominant driving force. The economist's basic mechanisms for stimulating innovation were absent, yet innovation there clearly was.

The major elements in agrarian change were new crops and crop management systems and an increase in the numbers of livestock. The novel crops, such as clovers, ryegrass, sainfoin and turnips, were all forage plants used for livestock feeding. They replaced field crops such as peas and beans. Grain prices remained static or declined after 1650 but livestock prices rose, so it is argued that a stimulus existed to grow fodder crops and increase animal output. It is also clear that there was a shift in emphasis from growing grain on heavy Midland clays towards growing fodder crops on the light soil uplands, where increased livestock numbers could maintain soil fertility. It is argued that after 1750 grain prices began to rise and cereal growing became more profitable, but there was no strong swing away from livestock for arable and livestock husbandry had become integrated by the growth of mixed farming over most of the country.[47]

An impartial observer might accuse economic historians of fitting a series of *ad hoc* hypotheses to their data to solve the particular historical problems of each period. Historians of the Tudors and Stuarts look at their period through different economic spectacles from those of the late seventeenth and eighteenth centuries (although there is more uniformity of approach between medieval and early modern historians). A student looking in detail at a set of technical procedures such as animal breeding must compose his account within these orthodox economic frameworks as he finds them, since he is in no position to formulate anything better. It is also true to say that the problems of identifying the forces which govern economic change are great and the arguments summarised here may have been oversimplified to the point of caricature, so that less than justice may have been done to their depth and quality.

We have so far dealt with the economics of agriculture in general. It is necessary to comment on the role of livestock husbandry within this general agricultural framework. Most of the crop and management innovations which occurred between 1500 and 1770, which must include the new forage crops, up and down or convertible husbandry, the floating of watermeadows and the ubiquitous enclosure and consolidation of open and common land, were all essentially designed to increase livestock numbers. This had the additional effect of prolonging and improving soil fertility so that grain yields could be maintained or increased on a reduced acreage. It has repeatedly been argued that the introduction of these innovations and the associated emphasis on livestock and mixed farming, compared with purely arable husbandry, could only occur in response to growing markets for livestock products. For this reason the conventional accounts of the widespread adoption of these new techniques place them firmly in the period 1650–1750. Before 1650, and especially in the sixteenth century, it is argued that the inflation of agricultural prices in relation to wages must have kept livestock husbandry and the incentive to improve it to a minimum since there was no mass market for luxury animal products.[48] By contrast, Kerridge has repeatedly argued that these innovations were adopted in the late sixteenth and early seventeenth centuries on a wide scale.[49] It certainly seems that the floating of watermeadows was adopted in the West Country on a fairly wide scale in the early seventeenth century.[50]

Whatever timing of agrarian innovation and the incentive to produce more livestock is accepted, the climate, topography and relative underpopulation meant that England in the sixteenth century must have been an essentially pastoral country. The two detailed regional farming maps and regional descriptions that have been published for the Tudor and Stuart period both show clearly the extent to which pastoral husbandry was dominant.[51] Even in the corn-growing regions the grain yield could not be maintained without sheep flocks to transfer fertility from the wastes to the arable.[52] But despite the importance of livestock in the rural economy of the whole country, writers on agriculture in the sixteenth and seventeenth centuries had very little to say on this subject, compared with their advice and exhortation to improve arable husbandry and introduce new crops and cropping techniques.[53] Their unspoken prejudices against pasture farming are summarised by Thirsk as follows:

They believed that most, if not all, land in the kingdom had once consisted of forest, and that by the efforts of man it had been transferred into pasture and cornland. This version of past history meant that the creation of cornland was the supreme end of the farmer's work. The arable farmer was always held superior to the pasture farmer.[54]

This is a pity since foreign commentators clearly believed the cattle and sheep of England to be superior to those of the Continent and admired the grazing skills necessary to achieve this superiority.[55] Tudor pamphleteers were also certain that the average English diet was far richer in animal products than that consumed on the Continent.[56] Nor must it be forgotten that many animal products were important industrial raw materials. Lubricants and artificial light were both provided from animal fats and oils, and enormous demands were made on leather for footwear, saddlery and industrial purposes.[57] Wool was necessary for the home clothing industry and although this remained static after the mid-sixteenth century there was nevertheless a constant, continuing demand. The argument that relatively depressed industrial wages held back both industrial and livestock product consumption is only valid if the industrial wage-earner was a representative consumer. But most of the population were engaged in farming with industrial by-employment. As farmers they benefited from increasing food prices even if they suffered from the rising prices of manufactures and increasing rents.[58] If large markets had not existed for meat, tallow and hides Harrison could not have written of the wealthy graziers of the late sixteenth century: 'Some such graziers are also reported to ride with velvet coats and chains of gold about them: and in their absence their wives will not let to supplie those tunes with no less skill than their husbands'.[59] The importance of animal husbandry in the established pastoral regions cannot be questioned, even if the agricultural texts of the day chose to ignore it.

It therefore seems reasonable to assume as a working hypothesis that a growing market for animals and animal products existed throughout the period from 1500 to the late eighteenth century which did provide incentive for innovation in livestock husbandry. Apart from those new crops and procedures already mentioned, it is clear that the livestock industry became geographically stratified and specialised over this period and that at the level of the individual animal there was a considerable drive for an increase in size. Later in the eighteenth century more sophisticated analysis of animal productivity shifted the emphasis away from mere volume of animal. The techniques employed by breeders to produce size increases and other animal improvements will be the main subjects of the chapters on different species which follow.

4

❖❖

The horse: breeding for war, sport and fashion

Introduction

What distinguished the breeding of horses from that of purely agricultural species was the existence of a fashionable and sporting demand among the aristocracy and gentry. The provision of these outstanding animals did not depend on purely productive or profitable considerations. The incentive was social competition and because it involved people with the leisure, money and education to devote time to the subject, the breeders of high-class horses had evolved many more guidelines to practice, and theoretical justification for them at a much earlier date, than had the breeders of either cattle or sheep. Considerable emphasis was placed on racing, hunting and coach horses in the seventeenth century but great aristocratic concern with horse breeding had also occurred in the sixteenth century. Here the object was the war horse since military tactics, at least in the early part of the century, were still derived from the medieval practice of deploying heavily armed members of the gentry on expensive chargers.

Much sixteenth-century horse-breeding theory was classical in origin, but it gradually developed into a body of thought based on contemporary observation and attitudes. The great agricultural livestock improvers of the later eighteenth century probably drew much of their practice and a great deal of theory from these earlier authors on equestrian matters and from the horse breeders themselves.[1] The breeding and trading of horses was an important industry in England in the sixteenth and seventeenth centuries. The growth in trade and industry produced a demand for more pack and cart horses and the transition from ox ploughing to horse ploughing generated an additional market for work horses. As the income of the rural population slowly increased, a new class of people appeared who used saddle horses or pads to get about, rather than going everywhere on foot. Sporadically throughout both centuries, there were military demands for cavalry remounts and draught horses for supply trains. During the sixteenth century particularly, the inroads made by military requirements created serious shortages of horses, especially in the southern counties, which were most accessible to the royal searchers.

The aristocracy and gentry had always required horses for hunting and hawking. At the beginning of this period these were usually the same horses that they used for riding and as cavalry mounts. During the sixteenth and increasingly in the seventeenth century, owners began to keep horses specifically for sport in greater numbers, not just for the traditional hunting and hawking, but for specialist equitation in imitation of the Italian riding schools (the art of the *manège*) and for racing at speed over level ground in matches and wagers. A new fashionable demand for heavier horses also began with the introduction of coaches and carriages, first for private use by the wealthy, and then as public carriages and stages patronised by the lesser gentry. The growth in the use of such transport after its first introduction in 1564 was very rapid and led to an extensive demand for matched and handsome draught horses, for which high prices were paid.

The history of this industry has hardly been investigated as yet, despite its obvious significance and its parallel with the motor industry of today. The techniques and theory of breeding have received little more than antiquarian attention and the early history of the horse race has not been considered at all.[2] In the sixteenth century the influence of the Italian school of horsemanship was very great in England and considerable government concern was expressed over the problem of the supply of suitable heavy cavalry horses in sufficient numbers. In the seventeenth century the influence of the Italian school waned as breeders and riders reacted against some of the more far-fetched features of their practice, while the nature of military and fashionable demand also changed.

Military concern throughout most of the sixteenth century, until effective firearms began to make heavily armoured cavalry obsolete, was with 'heavy' war horses, capable of carrying weight at reasonable speeds and having sufficient courage and stamina for mounted shock combat. During the seventeenth century the emphasis shifted away from such weight-carrying animals to horses able to move faster, with lighter loads, for much longer distances, as the nature of the deployment of cavalry changed. These horses were much more akin to the type thought desirable for hunting than the old war horse had been.[3] For both military and draught purposes high-class horse breeders made extensive use of foreign imports. In the sixteenth century most of these imports were from the Low Countries and Denmark, but later in that century and during the next, lighter-framed horses from Italy and Spain were used.

In the sixteenth century there was little mention of horse breeding specifically for sporting purposes. No breed was ever evolved in England specifically for hunting, and to this day good hunters, steeplechasers and show jumpers can be found among a variety of breeds and crosses. The situation was different with horse racing, for which the English evolved the Thoroughbred horse during the late seventeenth and early eighteenth centuries. Informal races and matches between horses and riders were an ancient institution, but there is considerable evidence of a growing fashion for, and standardisation of, racing procedure during the sixteenth century. Foreign horses were used extensively in the formation of the Thoroughbred in the seventeenth century, especially Turkish, Arabian and Barbary horses. In the sixteenth century all these types were comparatively rare compared with Italian, Spanish and Flemish horses, although the first two of the latter group undoubtedly had some North African blood in their ancestry. The royal and aristocratic stables contained a few Barbs and Turks, but most racehorses in the late sixteenth century were small, light, local saddle horses, usually from the peripheral regions of these islands, such as Irish Hobbies and Scottish Galloways, rather than foreign exotics.

The first clearly catalogued Arab horse in England was described in 1595 by Markham, who had trained the horse for its importer, Sir Francis Walsingham. The horse had come from the Angelica region of Arabia via the Turkish capital at Constantinople and then overland through the Balkans and Germany and finally to England by sea. So besotted was Markham by his knowledge of this single specimen that he advocated using an Arab stallion for the production of horses for every kind of work. However, his youthful enthusiasm soon wore off. In his book of 1607, Arab horses were not even mentioned.[4]

Escalating demand for horses continued unabated throughout the eighteenth century. The great majority of eighteenth-century animals, as in previous centuries, were literally 'workhorses' of no particular breed or merit, equine drudges of the plough, cart or gin.[5] The best draught horses from the breeding regions found their way into the stables of the gentry for coach work and into those of the wealthier urban traders as draught teams. The poorer ones, and those worn out from coach and the smarter waggon trade, went into the army for cavalry remounts and ordnance teams or were used by less fashion-conscious traders and manufacturers and by farmers who only required muscular effort from their animals. Less exalted saddle and road horses came from the same source, most working farmers and traders merely using their smaller and more lively draught animals for riding. The seventeenth-century market for superior saddle horses had largely disappeared in the eighteenth with the continued growth of private carriages and public post- and stage-coaches.[6] The specialised market for hunters was largely satisfied by the lighter sorts of farm work horse in the early eighteenth century. Only in Yorkshire and some of the other northern counties were horses still bred specifically for saddle purposes, although even in those regions where draught types were used for riding, some distinction was probably made between forms suitable for traction as opposed to the saddle.[7]

The history of all types of horse in the eighteenth century was modified by the development of the Thoroughbred, a breed of high fashion, great expense and enormous influence. In a sense the development of the Thoroughbred as a relatively fixed type by the mid-eighteenth century, substituted for Oriental and North African imports by making available in this country a variant of the Oriental horse, adapted to North European conditions. Oriental blood was now available on the doorstep in the form of cheap, second-grade blood horses, from which spirited cavalry, coach and hunting animals could be bred by crossing with humble, working brood mares.[8] The use of blood horses in many working breeds from the mid-eighteenth century was often overdone,[9] exactly as the use of the Dishley sheep was to prove counterproductive to many other sheep breeds later in the century. By the end of the century the use of Thoroughbred stallions in work horse breeds had settled down to a more sensible level, with the draught breeds maintained more carefully as pure varieties in their own right.

The leading horse breeders of the eighteenth century were therefore those concerned with the development of the Thoroughbred, those who bred draught horses that could be used for coach work, and those who bred the most fashionable team horses for the waggon or dray. The centre for the breeding of all three types lay in Yorkshire, Lincolnshire and Leicestershire and in the peripheral counties immediately to the north and south. The same general regions were also in the forefront of both sheep and cattle breeding in the eighteenth century, and the best explanation for the concentration of improved animal husbandry here must be that large tracts of these counties were not suitable for arable husbandry, nor were some of their soils good enough for the grazing of fatstock, so that the land had only low rent-earning potential.[10]

Fashion in the sixteenth century: the Italian school – cavalry and ceremonial horses

The sixteenth-century cavalry horse was the descendant of the medieval 'great' horse or Destrier, the horse which had carried heavily armed knights into battle. In the sixteenth century the type was generally referred to as a 'war' or 'trotting' horse.[11] Although difficult and expensive to maintain in the poverty-stricken societies of the late Dark Ages, these horses do not seem to have been particularly large by modern standards. They were certainly taller and more heavily built than other horses of their period, but all the illustrations, sculptures and descriptions of these animals from the time of Charlemagne to the sixteenth century suggest that the horses concerned resembled modern Cobs, whose main features are their moderate size, short legs, strong well-muscled bodies, good joints and feet. Their heads were also small and their necks usually short and frequently strongly arched. This Destrier in no sense resembled the modern Shire horse. That horse was developed for load-drawing purposes only and its slow pace, huge size, docility and soft bones render it completely unsuitable as a candidate for mounted shock combat.[12] But the originally noble Destrier was the basis for the development of medieval and sixteenth-century draught animals and there was considerable overlap in size and employment between the better waggon horse and cavalry horses, although this draught horse type did not resemble any of the modern heavy breeds any more than did the war horse of that period.[13] With the exception of these great horses, most animals were light, small and unsuitable for either draught or cavalry use.[14] The Tudors attempted to improve this situation by legislation and the extensive import of horses from the Continent, where there seemed to be types of horse more suited for military use.

Henry VIII's campaigns in the first half of the sixteenth century drained virtually all horses suitable for cavalry use from the south of England and a very large proportion of those from the northern counties as well.[15] The major responses by Henry's administration to the shortage were strengthening the existing bans on the export of horses out of England, the introduction of legislation to try and prevent small horses and mares from being kept on the commons and wastes, and forcing the aristocracy and gentry to fulfil their duties to the state for the provision of saddle and cavalry horses. There was also the carrot of advancement at court for a group of 50 Gentlemen Pensioners, whose major duty was the creation of studs of breeding mares and the supply of horses for ceremonial and military purposes. The king himself took the lead in making stud farms, and, through his extensive diplomatic and other contacts, imported stallions and mares from various parts of the Continent to try to improve the size and courage of English horses.

The original Tudor Act against the export of horses had been passed in 1495. Henry VIII attempted to strengthen the provisions of this Act and close any loopholes in various further Acts passed from 1531 onwards.[16] Further Acts were passed in subsequent Tudor reigns and the Privy Council repeatedly issued proclamations reminding citizens of their existence and sent instructions to the officers of the northern border towns and ports in an attempt to stop the export of horses into Scotland.[17] By the 1590s it was widely believed that the drainage of good cavalry horses from the north was entirely due to the open purchase by Scottish buyers at the northern fairs.[18]

The first Act laying specific obligation on the gentry was passed in 1536 and required the owners of deer parks to keep a specified number of mares not less than 13 or 13½ hands high, which should be covered by stallions of not less than 14 hands. Parks in the four northernmost counties were excused from this obligation, presumably because the Scots could all too easily get their hands on them.[19] In 1541 an Act with similar objectives was passed, but this time based on the wealth and position of social leaders. At the top of the scale dukes and archbishops were expected to maintain seven entire 'trotting' or war horses, with obligations decreasing with diminishing social and economic status.[20] When Elizabeth later issued proclamations to reinforce this statute, these included far more references to horses and geldings suitable for light horsemen, reflecting the shift in the emphasis of cavalry warfare as the century progressed.[21] In 1540 Henry enacted that horses kept for breeding on open commons, moors, forests, marshes and other wastes should be not less than 15 hands high at two years old in the majority of counties in England and Wales, and not less than 14 hands high in the northern counties.[22] This would seem to have been a forlorn piece of legislation, since Henry was demanding, in effect, that the smaller breeders and owners kept horses that were larger than those required of the upper classes. This fact was recognised during Elizabeth's reign when the obligation to keep 15-hand stallions was reduced to a requirement for 13-hand animals in much of Cambridgeshire, Huntingdonshire, Northamptonshire, Lincolnshire, Norfolk and Suffolk.[23]

Two important general points emerge here. First, it is clear that animals employed as war horses were in most cases perhaps 14 hands high, with a reasonable sprinkling of animals of 15 hands, while horses of 16 or 17 hands, approaching the size of modern Shire horses, were probably a rarity. Secondly, the legislators, who were also themselves keepers of studs of horses, were well aware that stature was predominantly an inherited character and that maintaining horses of sufficient size was a primary point in breeding. Whatever modifying effect the environment might be thought to have, inheritance was the dominant control over the form of the offspring of horses, a point which does not necessarily seem to have been believed for other domestic species, or indeed by some writers on equestrian matters.

Henry was also at the forefront in the import of foreign horses and, with perhaps equal significance, the import of foreign horsemen, particularly of Italians. Throughout the first half of the sixteenth century the undoubted leader among cavalry and equitation mounts was the aristocrat of Italian horses, the Courser of Naples. The term 'Courser' evidently came from the old French *corsiero* meaning a battle horse and the Italian Coursers of the sixteenth century were almost certainly therefore 'war' horses or chargers.

Subsequently, the word 'Courser' came to mean a swift running horse or racehorse, and considerable confusion has occurred in the literature because of the failure to realise that the term completely changed its meaning.[24]

As early as 1509 Henry had begun to ask for presents of horses from his diplomatic contacts.[25] Many of these imports found their way into the King's studs and the records of other members of the aristocracy and gentry show that such foreign stallions also featured in their breeding programmes.

When the Duke of Mantua sent over a gift of twelve brood mares in 1514 he also sent one of his Italian grooms and Henry was much taken with this gentleman.[26] So superior was Italian horsemanship at this period that royal and possibly other stables employed the services of Italian horsemen brought over for the purpose.[27] The select group of Gentlemen Pensioners were among the leaders of this cult of Italian equestrianism and, when printed texts on the Italian methods began to appear on the Continent, they were among the first to translate and annotate these works for English readers.[28] Both Thomas Blundeville and John Astley, whose works survive, were Pensioners, as was Sir Nicholas Arnold, whose manuscript on horsemanship seems to have been lost.[29] Gervaise Markham was the descendant of a Pensioner.[30] There also survives an Italian consultant's report on the state of the royal studs prepared by Prospero d'Osma in 1576.[31] The opinions of Blundeville, Markham, d'Osma, and the lesser work of Christopher Clifford,[32] provide a reasonable insight into the theory and practice of horse breeding in England in the later sixteenth century.

Since both royalty and aristocracy were importing foreign exotics into the country, Blundeville, Clifford and Markham all described these animals, some but not all of which were available in England at this time.[33] Blundeville's list was the most comprehensive. He outlined the following types:

(1) *The Turk*

He suggested that this referred to any horse from within the confines of the Turkish Empire which, since this included the Balkan States, Greece, Asia Minor, Palestine, Mesopotamia and Egypt, cannot have been a very homogeneous category. Turks had the reputation of being evil-mouthed, supposedly because they were not ridden until they were ten or twelve years old.

(2) *The Barbarian*

These would normally have been called Barbs and were brought in from the land of the 'King of Tunys' and from the North African hinterland behind it. They were small horses but they already had a high reputation for being 'verye swifte and able to make a very long carriere, which is the cause why we esteem them so muche'.

(3) *The Sardinian or Corsican*
These were very fierce horses and apparently very short in the body. There is little evidence that these animals ever played a role in England.

(4) *The Neapolitan Courser*
The famous charger and school horse of Italy was described by Blundeville as being strong, elegant and fast with a certain 'portlynes in his gait'. By the late sixteenth century it was no longer a great horse of the classic type but had been modified by crossing with light North African horses of the Barb type into a light cavalry or good saddle horse, ideal for the fancy control and riding of the Italian schoolmen.

(5) *The Spanish Jennet*
These were well-made riding horses. Blundeville described their gait as neither trot nor amble and was later criticised by the Duke of Newcastle, who asked the fair question of how on earth they did get about if they did neither one thing nor the other. It is probable that this high-class road horse also had North African blood in its ancestry.[34]

(6) *The Hungarian horse*
This was another saddle breed, noted for having the pace of a hard trotter.

(7) *Almayne and Flanders horses*
The largest of the classic war horse type, described as 'hard trotters' and already proving more useful as draught animals as the demand for old-fashioned heavy cavalry horses died away.

(8) *Friesland horses*
These were small, strong, compact horses of good stature, an excellent size for school work, but too stubborn and therefore unsuitable.[35]

Blundeville gave four main uses for high-bred horses. First there were cavalry horses (the trotting horse), secondly saddle horses (the ambling horse), together with swift runners and horses suitable for hunting and hawking. He recommended an ideal animal for each purpose which was a cross between two or more distinct types, a pattern that was also suggested by Markham, although their hybrid combinations did not necessarily agree.[36] Blundeville believed that the cavalry horse, which in his terms was still a 'great' horse, should be generated by crossing a Neapolitan stallion with a Flemish or native trotting mare. Possible alternatives for the stallion were German, Hungarian, Flemish or Friesland horses. Markham's ideal war horse was a lighter, more mobile charger produced from a cross involving English or Flemish horses on one side and Neapolitan, Turkish, Sardinian or Spanish ones on the other.

At first sight these recommendations to create animals for specific purposes by hybridisation seem to contradict the complaints made by a number of authors about the amount of indiscriminate crossing that had been going on to the great detriment of the horse stock of the country and in opposition to the basic principle, expounded by the Italian school, of the virtue of purity of breed.[37] Markham seemed to confound this principle even further when he suggested that the best crosses were often not merely the result of hybridising a horse and mare of different 'pure' breed, but the best stallions were those which were themselves the result of cross-breeding.[38] If such hybridisation had gone on in an uncontrolled manner for very long it would have been impossible to find any representatives of a 'pure' known strain at all, a state of affairs that the arch-cross-breeder, Blundeville, specifically complained of. This strongly suggests, although no sixteenth-century author explicitly stated this anywhere, that recourse was constantly had to 'pure' stock to recreate the specific combinations in each generation, in other words that the horse-breeding industry was stratified and that the stratification was not merely between breeders and users, but extended to the separation of pure base stocks from hybrid working stocks with desirable qualities combined from different purebreds, in the same way that the modern sheep industry is stratified. The maintenance of 'pure' stocks from which the ideal types could be synthesised by hybridisation was probably an objective of some breeders, although they were poorly organised and inefficient, and hence Blundeville's complaint. Certainly the majority of usable male saddle and hunting horses were gelded and therefore could not breed; consequently the type would have had to be constantly generated for users by breeders.

The authors offered much advice about the best geographical conditions for a stud farm, much of which had a distinctly classical flavour. Blundeville said that the specialist breeder should have extensive, well-organised grounds with plenty of small and large, well-fenced closes so that pasture might be frequently changed and various classes of stock kept apart from one another. The grass should not be especially rich, any such rank growth was to be kept for hay, while there should be trees for shelter, good water and some hard, stony ground for hardening the hoofs of the foals and strengthening their legs. D'Osma added that the soil should be dry so that dry grass would grow to counteract the naturally cold and phlegmatic disposition of mares. Rank and deep pastures merely reinforced their fluid spirits. Markham's objection to low ground and damp grass was that it encouraged disease, weakened the brood mares and gave weak-pasterned foals. Nor did his farm require the multiplicity of closes envisaged by Blundeville, because his breeding management was different. He suggested that three large closes would be adequate. Blundeville further believed in taming stud grounds as far as possible since he thought a domesticated environment would encourage tameness in the colts.[39]

Both Blundeville and d'Osma believed strongly that horsings should be as natural as possible with a minimum of human intervention. This opinion was echoed by Clifford, but opposed by Markham, who could see no valid reasons for not regulating the process for the practical convenience of the breeder. At a large stud Blundeville proposed that each stallion should have a close of its own, into which mares should be introduced in the breeding months of May and June, nine or ten mares for the lusty young stallions and fewer for the older ones. They were not to be allowed to run together longer than was necessary since the horse would be lost from work and the mares might cast their foals if they copulated after conception, an opinion which he had apparently gleaned from Aristotle. Even those breeders with only one stallion and a few mares should still run the stallion in a close, putting the mares in one at a time, a tactic which would overcome the objection of those, amongst whom was Markham, that the overexcited stallion would spread too much seed on a few mares and too little on the majority, thus encouraging cooler, more fluid conceptions and a superfluity of mare foals.[40] D'Osma went further and suggested the stallion should only be allowed to run with his mares in darkness and should be stabled by day. In his view a wild and spirited character in the foal could only be engendered in this way.

Both Blundeville and d'Osma abhorred the practice of standing the stallion in a stable or yard and bringing mares to him to be covered 'in hand'. D'Osma thought such matings would be dull and sluggish compared with the passion of a natural mating and therefore dull and sluggish foals would be the consequence, an aspect of the clear belief that the physiological and spiritual state of the parent animals at the time of copulation were critical to the quality of the offspring. He also thought, as did Blundeville, that such matings were usually sterile. Blundeville said that such a standing stallion would need very careful dieting and exercise to prevent him losing his natural heat, in which case excess filly foals would be produced. For the same reason, d'Osma would only tolerate this practice of 'in-hand' mating if the stallion was especially spirited.

Markham was a strong advocate of a 'house' or stable mating system on the grounds of ease of management and saving of space and fencing for closes. This consisted of leaving the stallion and mare together overnight in a large stable for a period of three or four nights. His theoretical justification was that the stallion distributed semen evenly among the mares if they were offered to him singly, and his practical one was that the kind of spirited stallion that he would ideally use could not easily be confined in a close. But even Markham's system was not so extreme and artificial as covering 'in hand', where the groom actually assisted the mating by bringing the horse and mare together under tight rein and encouraging the mounting and copulation by placing the 'yard' in the correct position in the mare.[41] Such practice has been standard in horse breeding from the eighteenth century onwards and was evidently practised in the sixteenth century although not approved of. Blundeville implied that he was shocked to find the tethering of mares to stakes and the covering of them 'in hand' was still used at the royal stud at Tutbury.[42] His attitude was quite clearly that such a technique was barbaric and unenlightened. But he condoned the practice of using worthless 'teaser' stallions to assess the readiness of the mares to mate, another procedure which had been used in antiquity.[43]

Whatever their disagreements about the value of natural mating, no writer doubted the value to the foal of suckling it as long as possible. Blundeville admitted that he was following classical precepts in suggesting that high-bred mares should only mate once every two or three years so that the dams might suckle their foals during this long period. Only lesser mares should produce a foal every year. He believed that something of the virtue (here largely meaning 'character' or 'constitution') of the mare would flow from her through her milk. Markham also recommended that the foals suckle for at least a year, while d'Osma suggested 18 months.[44] Markham felt the practice had the added advantage of ensuring the mare remained lean. He thought that overworked or fat mares would reject the seed and produce only small foals.[45] However, such long suckling periods and two-yearly matings were not standard English practice. Blundeville said that in general six months was the maximum suckling period, while d'Osma noted with disapproval that the same period (here five months) was all that was allowed even in the royal studs.[46] The theoretical belief in the advantage of maternal milk was clearly strong enough for the authors to think that the sacrifice in productivity would be worthwhile, but practical breeders evidently did not agree with them, and preferred to ensure a foal a year from their brood mares rather than invest too much capital in any one foal, however highly bred.

In classical theory, the age and physiological condition of the parents at the time of mating, and the nutrition received by parents and their offspring, strongly influenced the form and character of those offspring, since an animal consisted of the transposed elements of what it had eaten. Sixteenth-century authors held essentially the same views. Thus age of breedstock was thought important. Only Clifford believed the stallion should be as young and lusty as possible so that these characters would be transferred to his offspring. Markham, Blundeville and d'Osma all preferred maturity in the sire. Markham and Blundeville suggested the ideal ages were between 5 and 14 years, while d'Osma thought only horses of 13 years and upwards should be used as sires – after they had completed seven years of useful work. There were extenuating practical reasons for d'Osma's advice. He wanted to encourage the habit of keeping the stallions with the mares on the stud farms because he believed travelling them back and forth was detrimental to their generative properties. It is quite clear that the standard procedure of the time was to work the stallions and only bring them to stud at the breeding season. It was evidently not yet worthwhile for even the most important breeders to set aside stallions solely for procreation.

Contrary to his general advice, even Clifford thought the offspring of old stallions were not necessarily inferior,[47] and Markham conceded that one could use some 'principal rare horse' for breeding up to the age of 18 or 20 years without any danger of producing lame or diseased foals as some more classically orientated authors claimed. Blundeville, for instance, favoured the classical view that the defects of sloth and timidity would be transmitted by aged stallions and roundly condemned the practice in some quarters of only using worn-out, lame or blind horses for breeding.[48]

Whatever specific types of animal were used in breeding, whatever the nature of the grounds of the stud, however the animals were mated, and whatever the auspicious or unlucky astrological or imaginative conditions of the mating and pregnancy,[49] all the authors devoted most of their space to the principles of the selection of good parent stock, since without such favourable material none of the other influences on development could lead to high-quality offspring. Their recommendations on the phenotype to be used were often quite specific but generally combined with only vague mating and recording systems, partly because the organisation necessary to ensure tightly controlled breeding did not exist at this date, and partly because of the reservations most authors expressed over the 'morality' of close human supervision of the mating process. They were all firm believers in the hereditary principle that offspring generally resembled their parents in appearance, character and constitution.

Blundeville made the clearest sixteenth-century statement of the need to select parents carefully:

syth it is naturally given to every beast for the moste parte to engender hys lyke, as well in conditions as in shape of body, it is very requisite therefore for him that would have a good race, to be very circumspect in choosing the first stallions and mares from whom he would have his race to descend, to the intent he may reape thereof both pleasure and profit, and not bestow his cost and labour in vain.[50]

To ensure the continuation of the quality of the breed over the years (which seemed to be the highest objective he aspired to), the worst phenotypes – the sick, the old and the barren – which would inevitably occur in even the best studs through 'some evil aspect of the planets, or else by some unhappy chance, or the negligence of the keepers' should be regularly culled out to prevent them from breeding.[51] It would appear that within his breeding group, bearing in mind his reluctance to interfere with natural matings and his acceptance of the principle of allowing stallions to be working horses,

Blundeville's breeding strategy would have been barely positive, in the sense implied in Chapter 1. If the original parents were of different types, as he often recommended, Blundeville nowhere said whether the hybrids themselves should become breedstock, although he implied in his statement quoted above that the breeder would continue with a race descended from the first ones that he selected.

Markham's views on the matter seemed more clear-cut. He favoured the use of mixed stock, even the use of stallions which were themselves hybrid. But he did not hold with continuing with the same strain by interbreeding among one's own offspring, if, that is, it can be taken that his violent objecton to parent–offspring matings extended to matings among siblings and stable-mates, as it seemed to since he also opposed continuation in the same strain 'without any alteration or strangeness' on the grounds that this would lead to deterioration through inbreeding.[52] Combined together, Markham and Blundeville do seem to be saying that hybrid work horses should be re-created from base stocks in each generation and that those responsible for maintaining pure stocks should do so with the use of stallions of the same type from other places, so as to avoid the danger of incestuous relationships.

Markham and Blundeville were also the authors with most to say about the aspects of phenotype in the parents that were thought desirable, and they were in broad agreement about the majority of these features.[53] Apart from the age of the participants, the most important aspects of phenotype to be considered were beauty and constitution or character. The latter feature was obviously important for both horses and dogs, although of little significance for farm species, which perhaps explains the anthropomorphism which crept into discussions of horse and dog breeding. Both authors implied that the 'personality' was derived mainly from the sire, while the form and structure were inherited from the female.

The advice on shape and form derived wholly from the Italian school,[54] who in turn obtained most of their opinions from the classical canon of hippology. In addition, the Italian school had worked out a theoretical basis for colour preferences based on classical four-humour physiology. The theory depended on the belief that the colour of the coat of an animal was a reflection of the temperature of its constitution, so that coat colour markers, which were easily observable, were thought to be a guide to the all-important 'character' of an animal. Blundeville described the elements of the theory as follows:

For if he hath more of the Earth than of the rest [the other three elements], he is melancholy, heavy and faint-hearted, and of colour a black, russet, a bright or dark dun. But if he hath more of water, then he is phlegmatic, slow, dull and apt to lose

flesh, and of a colour most commonly milk white. If of the air, then he is sanguine, and therefore pleasant, nimble and of colour is most commonly a bay. And if of the fire, then he is choleric, and therefore light, whote and fiery, a sterer, and seldom of any great strength, and is wont to be of colour bright sorel. But when he doth participate in all the four elements, equally and in due proportion, then is he perfect, and most commonly shall be one of the colours following. That is to say, a brown bay, a dapple grey, a black full of silver hairs, a black like a moor or a fair roan, which kinds of horses are most commendable, most temperate, strongest and of gentlest nature.[55]
The argument developed along similar lines for all conceivable colour variants. Less obviously rational were the Italian precepts on the presence of blazes or marks on the feet, forehead and elsewhere, which were worked into an elaborate system of good and bad combinations. This marking theory seems to have been an imaginative fabrication without rational justification, although it may have been a smokescreen designed to conceal the real judgements employed by skilled horsemen.[56]

For both authors, the constitution or physiology of the horse was its most significant character, hence the attempt to find simple markers which would be a reliable guide to this inner feature, which was otherwise so difficult to recognise. Blundeville made the point that the behaviour of the schooled horse might not be a good guide to innate worth. Those elements such as carriage, obedience, responsiveness to the reins and so on, were 'artificial' and had been put there by schooling. A stallion should rather be judged by his 'natural' qualities, which Blundeville listed as health, fertility, strength, agility, swiftness, good disposition and aptness to be taught. This last would clearly have some influence on his 'artificial' capacities while some other characters, although basically 'natural', were inevitably modified by training, such as his speed over the ground and galloping action.[57] Blundeville here made a clear distinction between innate worth and the acquisition of value through environmental influence, one of the few instances where this difference may be said to have been clearly drawn by a sixteenth-century writer.

For Markham, the shape and colour of the stallion were no guide to his constitution. Experience had taught him that however good an animal looked it might well turn out to be worthless for sporting or military purposes because it lacked the necessary courage and stamina. Since appearance was such a poor guide to personality, Markham advocated that the ancestry of the animal be examined, in other words, that as far as possible in the days before stud books had been devised, the pedigree of the animal should be looked to. He drew a human parallel with the pedigree of gentlemen and aristocrats, suggesting once more that the origin of the pedigree-keeping system in horses was based on the principles and precepts of human family trees. In addition to beauty and conformation,

I would likewise have added unto them, his descent and generation; for albeit, a clown may beget a beautiful son, yet will he never beget an heroical spirit, but it will ever have some touch of baseness; and an ill-bred horse may beget a colt, which may have fair colour and shape, which we call beauty, yet still his inward parts may retain a secret wildness of disposition, which may be insufferable in breeding.[58]

Markham was a member of the gentry, and like the members and exponents of all hereditary castes from Plato onwards, he believed that his social position was justified not merely by historical accident and the possesson of wealth and power transferred from previous generations, but also by the transmission of hereditary biological qualities which provided a rational justification for his position. The obsession of all social hierarchies with their ancestry may be seen as an attempt to justify their position on the grounds of descent from some generally acknowledged notable or divine by an unbroken hereditary transfer (in the biological sense) of the qualities possessed by the foundation ancestor. Markham's anthropomorphic view of the qualities of a good horse led him to advocate the same basis of equine selection for precisely the same reasons of hereditary caste.[59] And Markham held this view despite (or perhaps because) the Elizabethan period had seen a great upheaval in the ruling classes, with the destruction of old elites and the creation of new ones. It cannot be an accident that formalised pedigrees and stud books were devised first for horses and hounds, which were bred for nobles and by nobles, and that the structure of these records was based on precisely the same premises as those used in human pedigrees.

Sufficient records of royal and noble breeding establishments have survived from the sixteenth century for us to judge whether the methods recommended by the equestrian authors were a reflection of actual practice. Henry VIII's own studs appear only vaguely in the records. In 1526 he had 32 parks at his disposal, and this number was greatly swollen between 1536 and 1539 by the dissolution of the monasteries. How many of them contained studs of mares is impossible to say. The French ambassador to England in 1542 said that he had seen two separate royal stables of 100 horses each and that Henry could draw on 150 new animals a year from his studs in Nottinghamshire and on the Welsh border. In 1519 he also had breeding studs in Warwickshire and Worcestershire.[60] However, it seems unlikely that he could have had establishments on a scale to provide 150 new horses annually. Certainly during Elizabeth's reign later in the century there were only two major royal studs – at Malmesbury in Wiltshire and at Tutbury and Hanbury in Staffordshire.[61] In 1546 a detailed inventory of royal horses was drawn up, probably of one of the working stables.[62] It contained 105 horses belonging to Henry and 10 belonging to various equestrian officials, confirming the reports of the French ambassador about the size of some of the stables. Of the king's horses, somewhere between 14 and 17 appeared to be pack and baggage animals, 30 were described as Coursers, cavalry horses in this context, and two as Moyles and 23 as Moylettes. These latter were Flemish horses and young stock from the same source, part of the army of Flemish horses in which Henry was increasingly interested during the 1540s. Twelve saddle horses were mentioned, described as Hobbies and geldings, and this stable also contained four exotics, Barbary horses from North Africa, a type whose import was recorded in royal correspondence in the 1540s. This period probably represented the earliest import of such animals. The stable also contained 8 stallions, entires kept for breeding, confirming that these animals were generally located at working stables.

Some detailed information about breeding policy may be drawn from the surviving lists of the Duchess of Suffolk's horses at Grimsthorpe in Lincolnshire in 1546 and 1547.[63] The 1546 list gave 90 horses and geldings of all ages and 35 mares, both stud and 'careage', the latter category presumably meaning working mares. They were divided into two types, 'trotting' or cavalry and 'ambling' or saddle horses. In the 1547 list, the majority of all horses were either stabled in Lincolnshire or London when the inventory was taken in April, and were evidently working, or potentially working horses. The brood or stud mares were carefully distinguished and listed separately as trotting or ambling mares. There were 6 mature ambling mares and 15 fully grown trotting mares, together with 3 more dispersed to the Marquis of Dorset and the Lord Admiral. Nine of the trotting mares were specifically described as Flemish, showing the extent to which the stables of the nobility were now using heavy Flemish mares as cavalry breedstock. The 15 trotting mares at Grimsthorpe were split into three groups: mares of the *Old Baye's*, *Lightmaker's* mares, and the *Corsair's* mares, of 5, 6 and 4 females respectively. The *Old Baye*, *Baye Lightmaker* and the *Cursare Grenadoe* appear as 3 of the horses held at the stable at Grimsthorpe and were presumably the sires of any of the foals born to the groups of mares designated to them. The sires used on the ambling mares were not recorded, suggesting their more humble status compared with exotically bred cavalry and display horses. The fact that the stallions were grouped with all the other horses in the stable indicates once again that they were not retained solely for breeding purposes. Some of the miscellaneous horses listed were described as 'of the king's stood' while others were said to be 'of the bred', presumably animals bred on the estate from Suffolk horses and mares. This also applied to those horses carrying *Suffolk* in their names. Other horses carried proper names and a colour, as in *Graye Sakford* for instance, which indicates the name of the family from which the animals were acquired.[64]

Of the five mares allocated to the *Old Baye*, four were Flemish and the fifth mare was described as the 'young Baye's mother' but given no origin. The

Young Baye was one of the horses listed at stable as a fully grown animal, the offspring of the *Old Baye* (from his name) and this last mare. This suggests a consistent breeding policy, where the same mares were allocated to the stallions for several consecutive seasons. By contrast, only one of *Lightmaker's* mares was described as Flemish. One was merely noted as *Baye*, two as *Young Mare Wase* and *Young Mare Corwine*, suggesting they were the offspring of mares of those names perhaps acquired outside the stud, the young mares being home-breds. The other two mares were called *Old Mare Boston* and *Young Mare Boston*, probably dam and offspring. Since it appears likely that mares were designated to stallions for several seasons, it is at least possible that *Young Boston* was bred out of *Old Boston*, with *Lightmaker* as sire. This could imply that *Lightmaker* was being bred to his own daughter, an incestuous mating of which neither Blundeville nor Markham would have approved. It would also seem that the Suffolks were owners of their own stallions and may have been 'continuing in their own breed', of which practice Markham would not have approved either. The four mares allocated to the *Courser Grenadoe* were all Flemish.

From his name, *Grenadoe* was undoubtedly a Courser of Naples. His mares were all Flemish, so that here at least the Suffolk trotting or cavalry horses were being bred according to the Italian precepts laid down by Blundeville. It is possible that the *Old Baye* was also an Italian horse, but there is no conclusive evidence. It seems unlikely that he was Flemish since he was not designated as such, so that, once again, the cavalry horses being bred from this group would also have been cross-breds. What of *Lightmaker?* His mares were not generally described as Flemish and were presumably English. It is possible that he was acquired by the Suffolk interests from a steelyard merchant of the same name who was acting as an agent for the hiring of mercenary horsemen in the Low Countries in the 1540s and may have traded in Flemish horses.[65] If this were so, *Lightmaker* was probably a Flemish stallion and since cross-breeding appeared to be the order of the day, he was not put to any Flemish mares.

Some of the animals bred in this stud were obviously carefully recorded, even if some of the young geldings and others destined to leave were not given any specific parentage apart from 'of the bred'. Those destined to join the breeding group, however, had their dams carefully noted. For instance, two year-old fillies were recorded as the offspring of *Old Wase* and *Young Wase* respectively. *Young Wase* was one of *Lightmaker's* mares and presumably *Old Wase* had been in the brood group and had now been pensioned off. If she had been one of *Lightmaker's* mares, another possible sire–daughter mating emerges. Since mares seem to have been so carefully allocated to stallions, noting the dam of any animal also recorded its sire. Such careful recording was a necessary prerequisite of any programme of individual selection or pedigree recording.[66]

Recording accuracy depended entirely on the calibre of the staff who were running the stud and stable. Harrison had said that Henry VIII's studs declined in quality as the officers 'waxed weary' of their task. Slack recording was still the order of the day at the royal stud at Tutbury in 1576 when d'Osma prepared his report on the state of the royal studs for the Earl of Leicester. His concern with accurate recording sprang from his advocacy of pure breeds. The purity of origin of any animal could not be judged unless its parentage was known. D'Osma did not suggest that individual records of parentage should be kept, but was content that the type of its parent should be known. Thus he suggested a system of only three marks, large for the progeny of large Coursers, a variant for the progeny of smaller Coursers and a middle-sized mark for the offspring of Jennets.[67] It was probable that there was no one of sufficient responsibility running the royal studs to allow for individual recording of offspring, such as had evidently been the case with the Duchess of Suffolk's cross-bred horses in Lincolnshire even though 'in-hand' mating, which occurred at Tutbury according to Blundeville, allowed more accurate records than the 'natural' system favoured by the Italians. Despite d'Osma's unfavourable comments on the practice of working the stallions at distant locations and only bringing them to the stud at the breeding season,[68] stallions were still not kept at Tutbury even when the stud was finally dispersed by Oliver Cromwell seventy years later.[69]

The park at Tutbury seemed to be large enough and divided into sufficient closes to allow for the natural mating system of the Italian school. In fact there was sufficient space for d'Osma to recommend that the Malmesbury stud be moved there too, since he disliked the soil and climate of Wiltshire.[70] There were at Tutbury, 20 Courser mares, 6 small Courser mares and 6 Spanish Jennets, the latter presumably kept to raise superior saddle horses. There were 43 young horses of all ages and sexes. Of the large Coursers, only 5 were to go to horse that year, the other 15, having foaled, were evidently to be rested for a year according to Italian teaching on the value of the dam to the foal. The 5 were to be covered by the grey Courser, *Grisone*, evidently an Italian horse, possibly a Courser of Naples. The small Coursers and Jennets were to go to the stallions, *Abbot* and *Argentine*, respectively. *Abbot* was so called because he was large, which suggested an element of cross-breeding by size, a practice to which d'Osma was theoretically opposed.

At Malmesbury there were 10 breeding Courser mares, 11 small Courser mares and 8 breeding Jennets. Only 31 colts and fillies of all ages were recorded, with no hint as to their type or parentage. No hint of the sires to be employed on the brood mares was given. However, 2 of the Courser mares seem to have had Courser sires, so they were presumably purebreds. By 1598 both Tutbury and Malmesbury could only muster 38 brood mares between them, 27 Courser mares, 4 young Courser mares and only 7 saddle mares for ambling geldings.[71] The royal studs at that date could not provide anything like the required numbers of replacements for the royal stables, even though these were probably more modest than they had been in Henry VIII's day. Sample inventories of the stables between the 12th and 33rd years of Elizabeth's reign show that the strength of 'double' (cavalry or state) horses and hackney or saddle horses varied between 229 and 261 head. In 1591 there were only 175 horses in the queen's stables.[72] In 1598 things had got so bad that the Tutbury stud only contributed 6 replacement three-year-olds, a far cry from Henry's reputed ability to provide 150 replacements a year from his own studs.[73]

Fashion in the seventeenth century: reaction against the Italian school – the *manège* Horse

Both equestrian literature and breeding practice in England in the latter half of the sixteenth century were thus dominated by the Italian school. Although elements of the Italian approach persisted into the seventeenth century, several treatises were composed in a spirit of intense reaction to the Italians and their classical models. Although written within the same philosophical framework and retaining much that was specifically Italian in origin, Markham's *Cavalarice* of 1607 already contained strong criticism of several aspects of the Italian system. The stimulus that pushed Markham into print was the publication of Edward Topsell's English translation of parts of the learned pandect of classical sources on natural history by Conrad Gesner,[74] which Topsell called *The Historie of Four-Footed Beasts*.[75] Gesner had translated all he could find in the classical authors on the subject of horses, and this same classical material was the basis from which the Italian school took their discussions of horse types and virtues. Whereas the Italians and their English followers supposedly described horse types as they were found in the sixteenth century, Gesner merely transcribed the classical material without reference to the very different form of contemporary animals. Markham therefore went through the various exotic horse types, pointing out where Gesner's descriptions were wrong, and incidentally finding himself in disagreement with Blundeville, who had sailed too close to the classical wind in his descriptions in a number of cases. Markham did not care for Gesner's descriptions of coat colours either, although he continued to accept the Italian theories of the reflection in colour of the character of the horse. It was his reading of Gesner which also induced Markham to reject much of the classical folklore about sex determination, which Blundeville had largely accepted, and to oppose the view that the offspring of aged stallions were necessarily lame or deformed. However, his rejection of classical advice remained highly selective and he continued to agree with Blundeville and the Italians on a great many points.

Much more radical in their critique of the Italian school were Nicholas

Morgan and William Cavendish. Morgan's work, *The Perfection of Horsemanship*, first appeared in 1609, just two years after *Cavalarice*, and was reissued in a second edition in 1620 as *The Horseman's Honour*. By profession, Morgan was a lawyer at the Inns of Chancery[76] and his attack was based wholly upon the philosophical basis of the Italian school and was highly theoretical. How much actual experience of horse breeding or management he had has not been recorded. The views of William Cavendish, Earl and later Duke of Newcastle, on horse management, schooling, riding and breeding first appeared in a French edition published at Antwerp in 1658 as *La Méthode nouvelle et invention extraordinaire de dresser les cheveaux*. This book had been composed by Cavendish while he languished on the Continent as a Royalist refugee during the Interregnum. After the Restoration, Cavendish was able to return to England and composed another book, similar but not identical to the first. The first edition of this work was published in 1667 as *A New Method and Extraordinary Invention to Dress Horses and Work them According to Nature by the Subtlety of Art*, with a second edition ten years later in 1677. Newcastle was perhaps the most famous and widely respected of all authors on equestrian matters in the English language in the seventeenth century and his influence stretched well into the 1700s. He composed his books in old age after a lifetime of successful training and breeding, so that most of his methods and ideas were worked out in the first half of the century. Newcastle's works were rather unsystematic and verged on the avuncular, very much the writing of a practical man rather than a theorist. Although he shared with Morgan a contempt for the attitudes of the Italian school, their reasons for disagreement with the Neapolitans were totally different. Cavendish did not believe their precepts because he had found from long experience that much of what they wrote was unsound. His interest in physiology or the theory of inheritance was negligible. Morgan, on the other hand, confined his discussions almost exclusively to inheritance and physiology so that, although he was able to point out inconsistencies in the theoretical base of the Italian school, the practical advice he gave as a substitute was itself based on theory just as arbitrary as that which he criticised.

Morgan stated explicitly in his 1609 edition that the Italians had had considerable influence in court and aristocratic circles in the late sixteenth century and mentioned the work of d'Osma in the 1570s.[77] Apparently d'Osma had soon afterwards fallen from grace and been dismissed from the royal service, and the direct influence of Italian horsemen in England seems to have ceased in the early 1580s. Morgan's whole book was composed in a prolix and tortuous style which obscured rather than clarified his meaning, so that it is difficult to be certain that one has understood his position and how he differed from the Italian school.

He began by stating that the Italians thought environment and schooling were the main determinants of quality in a horse rather than inheritance. Belief that the livestock of any region were wholly determined by local environmental conditions was certainly common among laymen, topographical writers and some agricultural authors[78] of the time and it was true that the Italians stressed the virtue of dressing or training a horse to make it into a good saddle or *manège* animal. But it would be unfair to suggest that the Italian school held such views. They were well aware of the importance of heredity in the selection of breedstock and Morgan was to a certain extent erecting a straw man here. In fact his initial statement on inheritance in breeding bears a striking resemblance to that of Blundeville quoted earlier. Morgan writes:

To begin then with the art of breeding of horses the first and principall foundation thereof consisteth in the election of stallions and mares, which are most fit and proper for that purpose, for the least carelessness therein is the utter ruine of all the whole work, since like engendring like, if any evil choice be made an evill product must remaine of the composition.[79]

But it would be fair to say that the gist of Morgan's discussion was to separate the effects of inherited from environmentally controlled factors in the breeding of stock. Authors from Greek times onwards, while conceding that offspring resembled their parents by virtue of what they inherited from them, were largely unable to separate a distinctly 'genotypic' component from the 'phenotypic', to distinguish innate from nutritional and other environmental influences. While Morgan did not come anywhere near making this modern distinction clear, he did attempt to show that the influence of the place where an animal was bred, the nutrition it received, and the age and condition of its parents were subservient to any innate contribution that it received from its parents, which was the essential information about the structure and form that the offspring should take.

Morgan's first criticism of the Italians was of their opinion that horses from different geographical areas were suitable for different purposes. He implied that much of this advice was based upon references to horses and regions in the ancient classical authors, the same point that Markham had made in *Cavalarice*. He suggested that this regional idea was rooted in the opinion that the worth of a horse was derived from the air, ground, water and climate of the region in which it was raised. Morgan thought this view false, although it could not be denied that some regions did produce better horses than others. He took as an example the case of northern and southern England, where horses from the former area were almost invariably better than those from the latter: 'Can this supposed excellency be taken from any other cause than from ayre, ground and water? Let us consider and examine whether this be an argument from no cause to a cause.'[80] His refutation of the influence of air, ground and water was based on the premise that form and condition of offspring were a consequence of uniting parental contributions and that this form was fixed from the first conception:

as he had from syer and dam, such will he be unto his death . . . for the best physicians and philosophers do hold that all creatures receive their conditions and qualities at the time of their framing and not at their birth, for otherwise nature were not perpetual.

Morgan suggested it was naive to suppose that real water was in some sense like pure 'elemental' water, of which the horse was partly composed, or that such impure, naturally occurring elements could alter the inner elemental qualities of the animal:

for if material water should alter the nature of the beast from the quality of his original creation, how should he be fit for the use of man, when necessity shall enforce him to drink all waters? and thereby have several alterations in quality, and therefore that being no principal cause there cannot be any effect of the alteration of original nature from the creation.

The same arguments applied to differences in ground and topography. Morgan was well aware that fens and marshes, for instance, were unhealthy places, but he believed that the ill-health resulting was accidental and did not fundamentally alter the nature of the person concerned since 'the alteration of nature hath not perpetuitie'.[81]

Another Italian notion was the system of coat colours and their supposed relation to the complexion or character of the horse, the result of the overall state of balance of the four humours in the animal. Morgan criticised the association between colour and complexion on four grounds:

(i) Since there were only four characters there should only be four coat colours.
(ii) The colours of horses altered year by year but their characters did not.
(iii) If the theory was true, multicoloured horses should have many characters, which was absurd.
(iv) If brown bay horses, regarded as owners of perfect characters, were as good as this system suggested, one would only have to consider colour in the purchase of a horse, but this was obviously not the case. Many other criteria were used by experienced horsemen.[82]

Thus he merely dismissed the Italian theory of the coloured marks or blazes out of hand as contrary to experience. Common observation showed that there were good and bad horses of every mark. He attacked Italian advice on conformation for its inconsistency, claiming there were some thirty different descriptions of what the perfect horse should look like. He went on to demonstrate to his own satisfaction that the features of shape specified were as likely to occur in jades as in good horses.[83]

At this point he believed he had refuted most of the advice of the Italian school, at whose door he laid the blame for the general predominance of worthless over worthy horses, an imbalance which he thought peculiar to his own generation.[84] Morgan believed that God was the original creator of both man and horse and that his original creations must have been perfect in every respect. Neither men nor horses were anywhere near perfect by Morgan's time, so he believed that considerable degeneration must have taken place. However, he thought the breed of horses could be restored to their original condition by searching out animals which possessed the qualities that had originally made them perfect, assuming that some examples still existed. This is one of the few seventeenth-century cases where it was thought that some positive change for the better was possible in livestock, not by breeding towards a type of animal 'designed' by man, but by breeding towards the recovery of past God-given perfection. In practice there can have been no difference in the results since the supposedly perfect original pattern was chosen by the breeder. Morgan asserted that there were six original primary qualities that horses had possessed. These he listed as boldness, lovingness, sure-going, easy-going, durability and free-going. In principle one should only breed from horses possessing these qualities to a good degree. The combination was in fact very much like that of the Italian school in that it depended on selection both for aspects of the personality or character of the horse and for suitable physiological features of its action.[85]

Since the nature of the horse was given to him by God at Creation and in subsequent generations this nature was imparted to the offspring at conception, this fundamental nature was not mutable, as were the qualities and characters acquired after birth. Critics might therefore object that if the nature of the horse was unchangeable, how had the degeneration occurred? Morgan marshalled the following defence of his position:

(i) The original Creation by God had been by divine means. Subsequent generations had been produced by natural processes, endowed with divine grace certainly, but of necessity less perfect. Hence there had been some accidental degeneration.

(ii) Using the authority of the Bible he argued that everything aged produced less perfect offspring than that which was in its prime. If the long passage of generations was considered as an ageing process from the first Creation, then the offspring were gradually deteriorating as each succeeding generation was further and further away from the origin.

(iii) Some argued that this ageing process through natural generation did not apply since the operation and development of the horse was caused by its soul, which was donated fresh at each generation by God. The fundamental nature of the horse had not degenerated with the passing generations, but the variation observed between good and bad horses was to be explained by the availability of elements that the soul was to use to form the actual horse, in other words that environmental influence controlled how they would turn out. Morgan conceded the justice of this point but implied that such variation from generation to generation would be essentially random, whereas he believed that the whole harmony of the elements had deteriorated in a directional sense.

(iv) Man had enhanced this deterioration by his breeding art. Only the perfect could produce perfect offspring. When breeders deliberately used old or worn-out animals for breeding, Morgan believed that degeneration would be enhanced since animals at an imperfect stage of their lives must produce substandard offspring.

(v) Some argued that degeneration could not occur to the species of horse as a whole, but only to particular horses in any generation as they experienced poor nutrition or conditions. Their procreative faculties were not subject to the same degeneration because these were concerned with the preservation of the species, which had not been corrupted since the Creation. But Morgan believed that the species was an artificial concept, that it was nothing more than the sum total of the individual horses and had no force, direction or destiny of its own independent of the separate component animals. Thus, if degeneration or 'corruption' was conceded on the individual scale, and no one could deny this, a complete corruption of the whole species over time would also have to be conceded since the imperfections of individuals were reflected in their offspring and procreation was not independent of the actual conditions of the animal as the Platonic concept of the species suggested.[86]

Morgan felt he had established two major points: (a) that the nature of the horse was primarily the consequence of its breeding and not of its nurture, while (b) the nature of horses in general had seriously degenerated since the Creation. His remedy was to select only horses which possessed good doses of the six fundamental qualities as parents, so that at least in those populations where his advice was followed there could be some general restoration of the better nature of the horse. His next problem was how to recognise these six qualities in individual horses. Using giant speculative steps, as arbitrary as any used by the Italian school, he attempted to find signs of those qualities in the outward morphology of the horse. Thus he suggested the following marks as indicators of a free-spirited horse: that it should have a slender, lean head and lean, thin and slender jaws. If horses of the conformation he suggested were used for breeding he believed considerable improvements in the general horse population would result. However, attention to the form of the parents in itself was insufficient to guarantee high-class offspring. To ensure the most perfect seed the parents should be at the right age, properly conditioned before copulation and should procreate under ideal conditions.[87]

Morgan thus conceded that the state of the seed was modifiable by the environment. While arguing fiercely against this as a factor dominating the development of the colt, he was forced to admit that it would influence parental seed and thus modify the progeny. This rather negated his original starting-point that geography was not a major influence on the form of animals growing in any region. But he stuck to his contention that the material seed was only modified by the environment and physiology and that it was not completely controlled by them.

On the age of horses he rehearsed the standard classical arguments for the middle years between five and ten being the best. Any combination of animals other than those of prime age would lead to excesses or deficiencies in some element which could produce an unfavourable complexion.[88] Similarly, nutrition would modify nature because the seed was drawn from the food and different foods clearly had different elemental properties. Thus, while the nutrition of the colt itself was held not to produce any permanent change in its nature, the nutrition of the parents producing the seed was thought to be significant. As evidence that the nutrition and work state of the parents were vital to the quality of the seed and the consequent conception, Morgan brought forward the fact that colts born from the same horse and mare were often different in their natures. Since they had inherited the same instinct and the same nature, the differences must have resulted from the different temperatures of the four elements in the parents at the time of conception[89] – an entirely reasonable attempt to account for the otherwise inexplicable variations produced by random genetic reassortment.

Morgan now admitted that many might argue from his advice that since nutrition did influence the resultant colt by the modification of the seed,

wherefore horses and mares, that are jades, being so dieted and kept, should not have a perfect temperature seed as well as the best horses and by consequent of the proposition as perfect and as good colts, if the perfection of generation consiste only in the temperature.[90]

He did concede that correct diet and management could greatly improve the seed of a jade, so that better offspring would result, but they would never be so good as those from an inherently superior animal since the jade was inferior by nature of the seed he had himself received at his conception 'otherwise he had not been a jade'. The nature of the seed was fundamental and would override any influence of nutrition or management, so that inherited quality was only modifiable within certain limits by nurture. But the converse was also true, that good horses incorrectly managed, could produce very inferior colts: 'And again I have found that a perfect horse and perfect mare may have a colt that is a jade, if my former rules be not observed.'[91]

In contrast to Morgan's recasting of philosophy, Newcastle's works were almost wholly devoid of any theoretical base. His opinions depended upon experience, not the ordered experience of scientific investigation, but the intuitive analysis of what he saw as he practised his craft. Newcastle had so little respect for natural philosophy that he merely dismissed those areas he regarded as nonsense, while accepting that which seemed to him reasonable. Although he mentioned most equestrian authors of the previous hundred years he made no reference to Morgan or showed any sign that he had ever heard of him or his opinions.[92] The abstract of Newcastle's views that follows has been prepared from the 1667 edition of the English work and a translation into English, with additions, of the chapters on breeding from the original Antwerp volume of 1658, prepared by Sir William Hope in 1746 as an addendum to his translation of the works of Newcastle's great French contemporary, Jacques de Solleysel.[93]

Like Morgan, Newcastle began by noting the great success of the Italian school in the latter half of the sixteenth century and then went on to dismiss both their systems of horse training and breeding. He used Grisone and his translator Blundeville as the literary representatives of the Italian school. He regarded as nonsense much folklore and many of the tales gleaned from the classical authors.[94] He also regarded it as nonsense to allow foals to run two years with their dams, a practice which cut their fertility and produced heavy and flabby foals into the bargain. He also rejected Varro's advice that the courage of a foal could be judged from its behaviour:

To know which fole will have the best spirit, by running foremost, and leaping of hedges and rails; is quite contrary to the experience I had once of a colt, that nothing could keep in, leaping over all things he came near; and when he came to be ridd, the dullest Jade that could be.[95]

While Morgan attacked the Italian system of colours and temperatures on its own ground by trading philosophical punches, Newcastle merely noted the common observation that good and bad horses came in all colours and shapes and any abstract theory employed to justify advice on this score was worth nothing in the light of experience.[96] He made the cogent point that in any case an entirely new theory of matter had appeared, the chemist's three-element theory of salt, sulphur and mercury. How could the colour system be justified on this new theory? Hope commented that Newcastle was wrong to dismiss colour from the guides to the quality of a horse, thus showing that some version of the system of a colour–quality relation was evidently still current in England even in the mid-eighteenth century.[97]

Newcastle concerned himself almost exclusively with school horses for the *manège* and said the only sure test of a horse for such purposes was to try him and see how he responded to control and how he could be made to move. Such a judgement could only be applied to an adult horse and the qualities sought were essentially internal, and therefore unknowable from its appearance or behaviour in the unschooled state. He did not favour Blundeville's point that 'good' horses could be assessed by their natural rather than by their schooled, artificial character. Newcastle's basis for the choice of a good breeding stallion therefore depended on selection for overall phenotype, where schooled behaviour was the major point of consideration. His preliminary advice was that a cheap stallion was a poor economy.[98] A 'good' expensive one would pay for itself in the superior value of its offspring. The problem, as always, was how to judge such a good stallion. Newcastle outlined the features to be sought as follows:

a superfluity of spirit and strength, docile and of an excellent disposition, and good nature, which is the chief thing in a stallion; for if he be of an ill disposition, vitious or melancholy, all his offsprings will participate of it, and will never be dressed, or made perfect horses as they should be.

He ought to be of a good colour, to give the Race a good dye; and well marked, to agree with most men's opinions: though marks and colours be nothing at all to know the goodness of a horse, nor shape neither; but the abundance of spirits and a strong chine, be the most considerable: Yet, by any means, I would have him perfectly shaped, for the beautifying of your race; for a handsome horse may be as good as an ill favoured horse; and an ill-favoured horse as good as a handsome horse.[99]

By contrast, his advice on the mare stressed her conformation almost to the exclusion of anything else, showing that Newcastle adhered to the general principle that the innate courage and 'personality' of a horse was derived from the male parent while the female contributed the physical structure. Only as an afterthought did he add that brood mares should be 'full of spirits and strength, and good natures'.

After getting the stallion to cover a mare twice in hand so that he understood what was required of him, he should be placed in a close with the mares he was to cover for six weeks. Here Newcastle followed the Italians in believing that natural matings were to be preferred to those in the stable, because 'Nature is wiser than Art' and because he thought fertility would be higher. Unlike the Italians, however, he suggested that in-foal mares could be placed safely with the stallion without any danger of seed or foal being cast. No more than ten or twelve mares should be allotted to him since the horse would herd them like a wild stallion and not eat while he was with them. If more mares were allocated and he had to remain longer with them to ensure that they all experienced oestrus, he would be too lean and weak to recover fully for his task in the following season, especially if he was worked throughout the rest of the year, as seemed to be conventional for stallions in the seventeenth century as much as it had been in the sixteenth.[100]

Newcastle gave specific advice about the building-up of a race of horses over several generations, a matter which Blundeville and Markham had left inconclusive. Like these two authors, he favoured crossing horses of different types in order to combine characters from several stocks. For instance, his favourite combination for *manège* horses was to cross a Spanish Jennet or Barb stallion with either English or small Dutch mares. He did not advocate re-creating the cross at every generation, but said the mares could be inbred to the same stallion over generations, so that sire–daughter crosses were permissible, although the stallion should always be obtained from outside your own race. Thus the mares, as the generations of the race went on, would become genetically more and more like the stallion type to which they were being mated, so that after three or four generations the mares would have been 'graded up', in modern terminology, to the breed of the stallion, with the result that the original cross, which had combined features from separate stocks, would be lost. However, it is doubtful whether Newcastle or any other contemporary breeder believed this transition would take place. The mares contributed only form, and the form was inherited through the female line, whatever the male cross employed. The mares, usually native, or of a type similar to the native, drew much of their virtue from the soil and environment of the stud, which was constant, and would therefore persist in the mares over the generations. Conversely, the recommended stallion was nearly always an exotic, whose character and personality would be diluted and altered in his male offspring by constant contact with local environmental conditions. The male once again provided the essential 'character' of the offspring, and to ensure a full dose of this, stallions from the original source must constantly be sought. It is doubtful, therefore, whether Newcastle would have found acceptable a Spanish stallion, bred from a Spanish stallion and mare in England:

I must tell you, that you must never have a stallion of your own breed, because they are too far removed from the purity and head, of the fountain, which is a pure Spanish horse; besides should the stallions be of your own breed, in 3 or 4 generations they would come to be carthorses so gross and ill favoured would they be; or at least, just such horses as are bred in that country, so soon will they degenerate: Therefore, have still a fresh Spanish horse for your stallion.

But you cannot breed better, than to breed to your own mares that you have bred; and let their fathers cover them; for there is no inceste in horses: and thus they are nearer, by a degree to the purity, since a fine horse got them, and the same fine horse covers them again.[101]

Newcastle thus accepted that the major control over the type and breed of any animal was the environment in which it lived, despite his recommendation of sire–daughter crosses to improve the 'purity' of sire blood in the second generation. He drew a direct analogy with human nationality, providing another example of anthropomorphic thinking among horse breeders:

For let a Frenchman remain in Germany, and his grandchild will be a true German; in like manner let a German live in France and his grandchild will be a Frenchman both in spirit and agility, such influence hath the climate, air and soil upon all creatures.[102]

Later in this passage he said that within a group of mares, selection of the best for breeding should take place. He clearly envisaged that there would be sufficient excess stock to allow for this, something that no previous author had made explicit. The selection among these mares might be towards the original form of their breed or, perhaps more likely, of the type which most closely resembled the sire. The consequence of breeding each new generation to foreign import stallions was to produce animals more and more like these imports, so that if a consistent policy of importing stallions always of the same type was maintained in one stud mare group, by the fourth generation, mares and horses produced from this combination would be on average genetically 95% of the import breed type. As stallions they would be almost indistinguishable from fresh imports. If at the same time some selection of type had gone on for a specific purpose, such as racing, the home-breds would probably perform better than the imported exotics, so there would no longer be any demand for them. Such a grading-up of native stock to Turkish/Arab and Barb horses would seem to have occurred in the case of the English Thoroughbred, so that by the mid-eighteenth century further import or use of oriental sires in the breed became counter-productive. The foundation of the official *Stud Book* in the late eighteenth century closed this type off from any new sources of genes and thus 'fixed' the breed as a type. Subsequently the pedigree information recorded in the books began to take on a mystique of its own, and breeders, seeking to overcome the lottery of genetic recombination in every generation, attempted to find some breeding method based on this information.

Apart from Spanish Jennets, Newcastle also had experience of Barb, Turk and Neapolitan horses,[103] in all of which he found a warm, masculine spirit, apparently absent from native stallions. He believed the most important feature of their native environment which moulded them was their nutrition. Therefore he offered the following advice about foals to encourage the persistence of exotic characteristics:

it matters not what kind of pasture they feed in, providing it be but dry . . . yae a man may bring up a very fine horse in his court; and that is the reason that the Barbs, Turks, Neapolitans and Spanish horses are so fine, nervous, so discharged of superfluous flesh, and so delicate a size, and so well proportioned, but only because they are brought up in a hot climate, and consequently with a dry kind of food? The secret then of bringing horses rightly up in cold countries, consists in nothing else, but keeping them warm in the winter, and feeding them with a dry kind of food, and in turning them out in summer to dry pastures.[104]

Newcastle discussed the main types of exotic horse known to him in more detail than any previous author since his knowledge was based not only upon horses in England, but those on the Continent as well. His main yardstick was the virtue of the animals as school horses for the *manège*; other possible functions he considered less important. His favourite stallions were the Jennet and the Barb. If the Spanish stallion were crossed with suitable mares, he believed it could provide good horses for schooling, cavalry, riding, hunting and racing. The Barb was himself a better horse for schooling than the Spaniard since he was less 'wise' or independent, but was not so fit a stallion to breed *manège* horses when crossed with native mares. In the original 1658 work he had stated that he generally preferred the Barb both as a horse and a stallion, but by 1667 the Barb was not so favoured, except as a sire for racehorses.[105]

He commented on the mixed state of English horse flesh, as had his sixteenth-century predecessors. English horses were no use for the *manège* compared with Barbs or Spanish crosses, but their mixed origin meant that some were sufficiently like foreign sires to be better than average. On the other hand, their mixed nature meant that they were the best general purpose horses available, most of them being reasonably useful for anything from school work to carting.[106] Friesland, Danish and Dutch horses were all heavier and less intelligent than English horses, but were adequate for school work. He noted that the fashion for Low Countries stallions, started when Henry VIII had discovered their virtues in the wars of the 1540s, had persisted, but that Dutch 'leaping' horses were increasingly hard to find by the mid-seventeenth century. So great was the demand and so lucrative the coach trade, that the Dutch castrated nearly all their horses for coach geldings. The shortage of entires meant that very high prices were paid for them: 'A town will join, and give above £200 for a stallion, but then he covers all the mares that belong to that town, like a town bull.'[107] A practice of which Newcastle seemed to disapprove since it implied in-hand matings and far more mares to one stallion that the ten or twelve that he thought appropriate.

Newcastle believed that wars in Naples and the whole of Italy had caused Italian breeds to decay (as indeed he thought English horses generally had declined through neglect during the Civil War and the Commonwealth). Neapolitan Coursers were now unimpressive. He had seen some that had been brought to Flanders at £300 a horse which were more suited to brewer's drays than saddle work.[108] However, it seems as likely that the Neapolitan was a victim of changing demand as much as decay in its breeding. Despite retaining some advocates, warfare with heavy cavalry was completely out of date in the mid-seventeenth century and Newcastle was firmly of the new school, favouring the lighter, nervous horse for both cavalry and *manège*. The famous Courser which Blundeville had so admired as a cavalry horse had been displaced. Dutch and Flemish breeders had begun selecting for the giant size which was to convert the 'great' war horse into a cart horse. As war horses, Dutch drays had already been rendered useless as Newcastle was aware. They had no speed or courage and their size was only possible through a lightening of their skeleton. He had compared the central space in the limb bones of a Barb and a Dutch horse and found that the Barb bone would scarcely accept a straw while the Dutch bone would take a finger.[109]

The most accessible stud records for the first half of the seventeenth century are the royal material from the reigns of James I and Charles I and those of James's favourite, George Villiers, Duke of Buckingham, Master of the Horse and equestrian enthusiast. In 1623 Villiers accompanied the then Prince Charles on his abortive mission to Spain to negotiate a marriage with the Infanta. Thirty-six Andalusian Jennets formed part of the leaving gift and Villiers later managed to charm many more from his contacts in Spain. The high fashion for Spanish horses began from this period. These records give some hint of contemporary breeding practice, which can be compared with Newcastle's literary advice.

Prior has transcribed most of the inventories and other extant data about the royal and Villiers studs together with an actual horsing record that has survived from the leading breeder, Conyers Lord d'Arcy, at Hornby in Yorkshire in the early seventeenth century. Both the d'Arcy and Villiers records were unusual in the large number of stallions used on a small number of mares, so that many stallions only covered one, two or three mares. This may be a distortion since the d'Arcy record was a mere jotting and the Villiers horses at Hampton Court were perhaps only a few which happened to be in the royal stables and park there, or conversely, it may represent an active policy of experimental crosses among horses of different types. The royal records of the 1620s were more conventional in that the minimum allocation of mares to stallions was four while the majority of them were given six or more.

The d'Arcy note at Hornby recorded only five brood mares in each of two years, 1617 and 1620.[110] Three of them were the same in both lists, suggesting a policy of retaining a small group of brood mares for a reasonable length of time. The two new mares in 1620 were *Grey Royal* and *Grey George*, indicating that they had come from the royal stud and George

Villiers's stud respectively. Four stallions were used on the mares, with one stallion in each case covering two mares. In both years two of the stallions were Barbs, different ones in each case. The names of the others carried no indication of foreign origin, nor did any of the mares. Since exotics were often noted as such, it seems reasonable to assume that animals given simple names were native or native–cross horses and mares. One of the mares common to both lists was *Brown Spink*, while in the 1620 list one of the stallions was called *Bald Spink*. Horses carrying the same name were usually related and it is conceivable that *Bald Spink* was out of *Brown Spink*, so that d'Arcy was perhaps ignoring the advice of nearly all authors that home-bred stallions should not be used. In 1620 the mare, *Grey George*, was put to the *White Barbary* twice and on the second occasion d'Arcy noted what he assumed was an auspicious sign, 'the moon being new'. He accepted that astrological influences had some part to play in breeding.

In Villiers's covering lists at Hampton Court, 13 mares and 6 stallions were involved in 1620 and 21 mares and 8 stallions in 1623.[111] Many of them were clearly designated foreign while others which were named after, and therefore acquired from, individuals with strong Spanish connections may have been Jennets.[112] One stallion in the 1620 list was called the *White Courser*, and a horse in 1623 was called the *Black Courser*, which suggests they were both of Italian origin. Of the six sires employed in 1620, three were Barbs, one was a Jennet and two were Coursers, probably Italian. Three of the mares were described as Barbs, four as Spanish and the other six carried no indication of their origin and may therefore have been native or part-native mares. The *White Courser* was run with his three Spanish mares in the park for six weeks in a 'natural' mating, but for the others precise covering dates were given, suggesting that the majority were covered in yard or stable and possibly in hand. The balance of stallions in 1623 had changed, reflecting Villiers's close connections with Spain at that date. Two were Barbs, one was a black Courser, while the remaining five were Spanish Jennets. Of the mares, two were specifically called Barbs, four were designated Spanish, while the remaining fifteen had no indication as to origin.

Only a minority of the matings were designed to continue a pure breed, five to perpetuate Barbs and four to continue the Spanish Jennets, a total of 27% of matings from both years. The crossing pattern for 1620 otherwise looks like a random assortment, while the 1623 list was dominated by the cross of imported Andalusian Jennets on native mares. This last cross was one that Blundeville recommended for saddle horses and was to be Newcastle's favourite combination for schooled saddle horses. He may well have learned its virtues from the royal stables, where Andalusian horses were present in profusion.

The three covering records from the royal studs at Malmesbury and Tutbury in the 1620s show a reasonably consistent breeding pattern.[113] In contrast to the Villiers arrangements, a smaller group of stallions was used. Thus in 1620 at Malmesbury five stallions served 39 mares (8 mares to each stallion on average), while in 1624 at Tutbury six stallions covered 47 brood mares (a very similar ratio of almost 8 mares to each stallion). As was by now conventional in all high-ranking studs, all the stallions were foreign. At Malmesbury three of them were light horses, one Turk, one Barb and one Jennet, while the other two were probably heavier, designated the *Ambling Courser* and *Bay Poland*. At Tutbury there were three light southern stallions, one Barb, one Jennet and a very young horse referred to as the *Arabian Colt*. The other three appear to have been heavier horses, an ambling *Courser Digby*, a grey Courser called *Potts* and a white French horse given the same name. It is clear that despite Villiers's influence and enthusiasm for Spanish and Mediterranean horses, they had not taken over the royal stables on the sire side. This is not surprising since the purpose for which the studs were maintained was still to provide state horses for formal occasions, together with high-class cavalry mounts, not hunters, racers or *manège* horses for which Jennets and Barbs were recommended. Only three of the 39 mares in the Malmesbury list were called *George* and can therefore definitely be said to derive from Villiers, while in 1624 at Tutbury, somewhat surprisingly, there was no horse called *George*.

A very large proportion of the mares were of foreign extraction. In 1620 at Malmesbury, definite Spanish and Mediterranean mares numbered ten: four Turks, four Jennets and two Arabs, with the three *Georges* also perhaps of Spanish origin, while ten were from Savoy, the Low Countries, Germany and North Central Europe and were probably heavier types. There was one Courser, perhaps of Italian origin, giving a total of 24 from 39 mares which appear to have been foreign. These mares, like the stallions, seem to have been reasonably balanced between the heavier, northern and the lighter, Mediterranean types. Matings among these types do not seem to have followed any consistent pattern. Every conceivable cross occurred with no particular preference given for either Blundeville's Courser x Low Countries cavalry horse or Newcastle's Barb or Spanish x English or Dutch mare cross for a schooled saddle horse. The 47 Tutbury mares of 1624 were not so clearly named. Only three were specifically called Jennet or Barb, although the two called *Don Juan* may have been Jennets, while four were referred to as *Courser* and another six as *Savoy*. However, the 1624 list does provide good evidence that the brood mares in the royal race were being managed as a closed breeding group, as Newcastle advocated, with the stallions coming from elsewhere and not from the breed of the royal mares.

Large groups of mares in a stud with the same name suggest a policy of breeding replacement mares from among your own race, rather than bringing them in by gift or purchase. Both Blundeville and Newcastle had said that once an initial group of brood mares had been selected, it was better to persist with this group and their offspring rather than constantly chop and change the brood group. The Malmesbury list of 1620 provided a hint of this policy as there were five mares called *Vadomon* and two called *Hunsdon*. But in the 1624 list the multiple naming became much more prominent.

Apart from the six *Savoys* and the two *Don Juans*, there were eight *Polonias*, three *Truggs*, two each called *Brilladore* and *Emperor*, and three each called *Burley* and *Douglas*, of more doubtful provenance since these came from outside the stud, from the families named. At this time, if Newcastle may be taken as accurate, anything resembling a pedigree breeding system had not yet evolved. But within any given stud policy decisions about breeding could be more rationally taken if the parentage of all the foals born was carefully recorded. If they were all derived from a brood group which remained fairly inbred once the original animals had been selected, the record of the parentage on the female side was adequately dealt with by the system of adopting the same name. However, the stallions were usually from an outside source and the sire of any brood mare born within the brood group and used as a replacement would need to be carefully noted. In the stud list of the Duchess of Suffolk in the sixteenth century, this was not done as carefully as it might have been, but in the royal studs the sire of each foal was carefully monitored, even when these animals were destined to come out for a life of work in the stables. Thus the seven remaining three-year-old colts born in 1617 which were ready to come up to the stables from Malmesbury in 1620 each had their sire carefully designated. The same care can be seen in the stock list prepared at Tutbury in 1649 just before the stud was dispersed, where the sire of every brood mare, filly and colt was clearly given. This information may by then have been of interest to potential buyers in their attempt to judge the quality of what was on offer. Once a sufficient number of breeders had begun to keep adequate records, the pedigree of a horse could be passed on to its next owner and the beginning of a pedigree 'system' of breeding could be said to have occurred.

As far as the royal studs were concerned it would seem they had to ensure a reasonable supply of riding and state horses for the king and an adequate supply of replacement brood mares for the stud as the older mares became unfit or barren. Analysis of the records suggests that they were probably incapable of meeting likely royal demand.[114] The mares at Tutbury and Malmesbury were undoubtedly superior to those likely to be found in the general run of gentlemen's studs, but they were not that much more valuable than some of the better-class animals owned by the gentry or indeed those available at the most notable horse fairs from dealers and coursers.[115] The dispersal of horses and mares from the stud in 1649 was not recorded, so that most of the buyers cannot now be traced. One has to rely on stray references to 'Tedbury' horses and mares in contemporary archives, as for instance the purchase by the Kentish gentleman, James Master, of a stoned horse of that breed in 1653 for £26 and another in the same county in the possession of Sir Thomas Pelham. These were almost certainly some of the horse colts recorded in the 1649 list,[116] not the stallions, which did not stand at the stud and were probably not of the king's ownership for the most part.

So much for the first half of the seventeenth century. Later, and during the eighteenth century, the whole development of horse breeding was strongly influenced by the rise of the Thoroughbred. This breed and its origin really deserve a separate chapter, but instead it will receive only the lion's share of discussion of eighteenth-century events.

5

Horse breeding in the eighteenth century: blood, speed and carriages

Horse racing and the origin of the Thoroughbred

The modern form of flat horse racing over relatively short distances evolved during the eighteenth century but references to horse 'chases' and 'races' are of much greater antiquity. Originally these were hunting contests, but during the sixteenth century the controlled race or match between animals over level ground became more popular. Markham described both activities in some detail. The 'chase' had arisen from the ancient practice of hunting wagers, where gentlemen had bet between them on how well their horses could keep up with a full day's hunting. The late sixteenth-century derivative was called the 'wild goose chase' and by this time specific horses were being kept for this sport. It was disapproved of since the hard riding in dangerous conditions led to the deaths of many animals. The object was to test the speed, stamina and skill of the horse and rider. One horse was designated leader, the other follower. The lead rider had to retain his lead by hard riding and various devious stratagems to prevent his being overtaken by the follower, who, in turn, was not to fall more than a certain distance behind. The chase continued until the leader lost his lead or the follower fell more than the prescribed distance back. The wild dash had originated as a means of providing a sudden death play-off when both wagered horses had kept up equally well with the day's hunt. By the mid-sixteenth century these hunting wagers frequently took place with the hounds following the trail of a dead cat to prevent checks and keep the horses flat out all the time. By the late sixteenth century, the wild goose chase had become an end in itself and was no longer necessarily associated with a hunt.[1]

But matches of coursing or racing horses over shorter distances of open ground as tests of pure speed were also common by this time and Markham gave as much space to the dieting and training of racers as he did to hunters and chasers.[2] The type of horse required for hunting or the wild goose chase would not be quite so suitable for flat coursing (here used in the sense of

93

'racing'), although the same type of horse would still be employed at this date and no specialised racing horse breed had yet evolved. Markham implied that a horse regarded as faulty for hunting was probably adequate for racing and therefore coursing horses were probably reject hunters, something of a reversal of modern practice where second-rate Thorough-breds often wind up as hunters, steeplechasers and show jumpers.[3]

The Earl of Rutland's horses used for matching at the very end of the sixteenth century were referred to alternatively as 'Hobbies' and galloping horses, which suggests that adequate racehorses could also be found among the ranks of fast saddle horses.[4] Rutland had his horses specifically in training as runners and they were raced regularly at Lincoln and Royston between 1602 and 1615.[5] In 1608 the fastest horse in Yorkshire was alleged to have been worth only £20 before being put into training, while conversely in 1618 a horse was in training for a race at Rugford Hill in Lincolnshire which was said to have cost £140.[6] By the first half of the seventeenth century Newmarket Heath was becoming a major site for high-class racing. The Earl of Salisbury kept a racehorse, *Rainbow*, permanently in training there between 1640 and 1644.[7] At about the same time the first Earl of Pembroke owned a string of racers, of which two, *Peacock* and *Delavell*, were nationally famous.[8]

After the Restoration there was a quantum jump in the development of 'professional' horse racing. Plate and other objects were still raced for, but now the value of the plate was the incentive, rather than the honour of the trophy. At the beginning of the century the same trophy had been run for year by year, but in the 1660s the trophies were kept by the winners and plate for the races was bought by subscription each year.[9] The rules and eligibility of horses for the races became more precise and were stratified into courses for horses of different ages and classes, the beginning of the series of graded trials that racers had to go through before they could join the elite for the most valuable and prestigious events.[10] The money wagered on matches became much more serious at the same time. Lord Roos owned a racehorse, *Bo Peep*, which won £1000 in a single match at Newmarket in 1667, the stake shares apparently being for 50 guineas.[11] By the 1690s a mature system and season of racing had evolved with other famous training grounds apart from Newmarket, such as Banstead Downs. The three most famous meeting grounds at the end of the seventeenth century were Newmarket, Burford in Oxfordshire and Quainton in Buckinghamshire. The Rutlands also attended meetings at Farnden, Chester, Penkridge, Lincoln, Woburn and Tiddeswall in 1696 and 1697.[12] John Hervey, Lord Bristol, won 1170 guineas on a match in 1698 and a specialised elite of racehorses had certainly evolved by this date.[13]

Some of these horses were of 'blood' or Thoroughbred type and the story of

Plate 5. The match between *Aaron* and *Driver* at Maidenhead, painted by Richard Roper (*c.* 1730–1775). The two horses are shown finishing the third of three matches or 'heats' run in 1754. Such a match series between two horses over flat turf was a linear descendant of the hunting match and 'wild goose chase' of the sixteenth and seventeenth centuries. The true race or course where a field of horses competed was increasingly common by this date. See J. Egerton, *British sporting and animal paintings, 1655–1867* (Paul Mellon Collection, 1978), p. 110. Photograph from the Tate Gallery, London.

flat racing in the eighteenth century centres around the evolution of this breed, although it is clear that until the late 1700s the Thoroughbred did not completely dominate the racecourse. The development of this breed and of the other types of horses raced remains obscure, but it is possible to indulge in some reasonably well-informed speculation about the breed and its precursors.

The English Thoroughbred has become the archetypal flat-racing animal throughout the world and its massively recorded pedigree information, largely accurate from the 1790s to the present day, has led many to believe that the breed was deliberately created as a racing machine and the secret of its racing value must lie somewhere in this genealogical information. But this completely misunderstands the virtues of the Thoroughbred and the reasons for the publication of stud records. Entry into the canon of the Thoroughbred, as in the case of any closed breed or race, was dependent entirely upon parentage and no other criterion; were the parents of the animal concerned for inclusion within the breed records themselves pure-bred examples of the breed?[14] The only way to judge this was to know the parentage of these parents in turn and so on back to a foundation group which were probably a purely arbitrary assemblage, since the need to compile accurate breed records did not usually arise until well after the breed had emerged as a coherent entity, so that records of the foundation stock were already historical data when they were compiled. The stimulus for the public collection of pedigrees in the *General Stud Book* at the end of the eighteenth century was the identification of horses and the accurate assessment of their age, since different classes of race and prize existed for horses of different ages. Checking on the *bona fides* of horses entered for races was becoming both expensive and tedious late in the century. The temptation to cheat on the identity or age of an animal became greater as the prize money for races increased and as the races with the most prestige became confined to younger and younger horses.[15] So much money (and pride) was at stake that a published record of the parentage, foaling date and a detailed description of each animal, together with an effective policing body, the Jockey Club, to enforce the rules, was the only way to organise a complex racing calendar fairly and effectively.

The use of exotic stallions had been obligatory among breeders of all elite horses for whatever purpose since the sixteenth century at least and racing was no exception. The emphasis, as we have seen, was on matching up the characters of animals from different origins by hybridisation between local and imported breeds. Most authors believed that the virtue of horses from exotic locations was only transmissible over generations while they remained in those places. The Duke of Newcastle reported that Sir John Fenwick, a leading breeder of 'running' horses in the 1620s and 1630s, had always used a Barb stallion on local English mares, choosing a Barb as close in conformation to his mares as possible.[16] The half-bred, or sometimes quarter-bred, Barb racers were not then used by Fenwick as stallions. He said that he went back to another Barb import so that he would continually re-create the Barb-cross type in each generation. This approach agreed entirely with Newcastle's own view on the need to revert to the fountain head on the stallion side at each generation. Since Fenwick sought mares near to the Barb type, it is probable that he used animals which were themselves Barb x English crosses and that his mares became more and more loaded with North African genes – if he adopted the recommended policy of home-breeding of replacement mares. Provided that his stud mare collection was not dispersed (which most studs normally were on the death or fall from grace of their owners) his race of horses would grade more and more towards the Barb type if a consistent breeding policy was persisted with for long enough. These horses of Sir John's were not Thoroughbreds because no attempt was made to create a closed group of pedigreed animals. He undoubtedly maintained records of his breeding mares but he would not have accumulated their pedigrees because there was no part of his breeding objectives which demanded that he should do so.

After the Restoration it is probably safe to assume that quality horse breeding had to begin again more or less from scratch and elite horse breeders seem to have turned almost exclusively to the generation of racehorses. The type of horse required for the races and matches of the last forty years of the seventeenth century and the early part of the eighteenth was not the type of the modern, sprinting Thoroughbred. All the horses matched for trials in the earlier period were mature and at least six years old, and at the beginning of the eighteenth century they were not infrequently run over distances of six or even eight miles, although such distances became rarer as the century progressed, the usual match distance for mature animals becoming four miles. The weights carried were usually between eight and ten stone, although in the King's Plates – proper races rather than matches or trials – the weight carried was usually twelve stone since the purpose of these royally endowed races (instigated by Charles II and continued by all monarchs until George II) was to improve the stamina and military capacity of saddle horses, an indication that racing, as well as hunting and chasing, had originally been introduced as much for military training as for purely sporting or fashionable reasons.

Prior stated that the first race to be run exclusively for immature horses at Newmarket was a race for four-year-olds instigated in 1730, while from about 1750 onwards weight-for-age handicap races became common, with all ages of horses competing down to four-year-olds, but carrying proportionately less weight than fully grown seven-year-olds and upwards. He also stated that the first £50 Plate for three-year-old horses was not run at Newmarket until 1756, that there were no two-year-old races until 1769 and that throughout this period and the 1770s the racing of such young animals was not popular – the Newmarket Sweepstakes for two-year-olds in 1775, for instance, attracted only three entrants. Set against this, however, is the fact that Lord Bristol was entering a colt of his for a 100 guinea Plate at Newmarket in 1712 in a race specifically for colts, probably either two- or three-year-olds. It may be that races for young horses were thought of as colt races in the early part of the century and not entered in the official racing calendars and that the transition to races with young horses had been going on for some time without the official records, such as they were, showing it. The time seems rather short between the feebly patronised two-year-old Sweepstake of 1775 and the establishment of a 'classic' two-year-old race, the July Stakes, at Newmarket in 1786. It looks even more surprising since the three classic three-year-old races over distances up to $1\frac{3}{4}$ miles were established in the 1770s – the St Leger in 1776, the Oaks in 1779 and the Derby in 1780. These firmly established the modern primary form of the flat horse race as a sprint for young horses carrying weights of about eight stone. According to the Duke of Grafton in 1784, Thoroughbred horses only raced between the ages of three and six, which again suggests that the transition to racing young horses was probably not so late or so dramatic as Prior at least seemed to believe.[17]

Nevertheless, the change in the form of race from a four-mile course for mature six-year-olds and above carrying weights up to twelve stone, to sprints of less than two miles with horses of three years old carrying weights of eight stone, was dramatic, even if the change was slow and occupied much of the eighteenth century. It seems to correlate with two other phenomena: the eclipse of imported Oriental and North African stock as desirable material for siring racers and the origin of concern with the Thoroughbred as the exclusive form of racehorse. While the origins of the breed certainly go back to the mid-seventeenth century, it only became pre-eminent on the racecourse once the style of racing had changed to suit its characteristic, brilliant sprinting capacity, making other breeds and crosses less suitable for racing purposes. Having established itself, it became a closed breed from the mid-eighteenth century. The eclipse of the Arabs and Barbs merely reflected the decline of the traditional cross-bred distance racers, which had required weight and stamina to stay the very long distances run. Arabs, Turks and Barbs themselves had seldom made good racers under these distance conditions, and by the time that sprint races became fashionable a local variant of the Oriental type, the Thoroughbred, had been developed which was adapted to local conditions and could easily outperform any Mediterranean import that was put into training.[18]

The essential character of the Thoroughbred has probably changed very little since the late eighteenth century because the races for which it is suited have not changed and breeding techniques that rely upon the analysis of pedigree, and the unsatisfactory nature of the horse race as a controlled test of performance, have probably been of little selective value in altering the racing capacity of the breed. There have been few changes in fashion or the market, such as have occurred with farm livestock, to induce any major change in selective pressure or breeding technique, and it is possible that the breed has ossified since the late eighteenth century despite the claims of racing men that the breed has been improved continuously since then.[19] What evidence can be produced to support the view outlined here that racing has altered to suit the Thoroughbred, rather than the breed having been designed for a specific type of race, and what can be said about the origin of the breed in the first place?

What did the term 'Thoroughbred' mean? It was seldom used in the eighteenth century and was not mentioned in the first edition of the *General Stud Book*, although there is a record of its use as early as 1713. It seems possible that the breeders of this type of horse, whether they called it a Thoroughbred or not, were attempting to maintain the purity of imported Oriental stock from Arabia, Turkey and North Africa, between which types very little distinction was made,[20] although subsequent research has shown that a considerable number of diverse and unrelated strains were involved among the animals imported. The only guarantee of the membership of offspring in each generation to the breed type concerned was their pedigree, so that among the relatively small group of breeders who tried to maintain this purity of descent the detailed pedigree of the animals became very significant. Their major, and perhaps only, objective was the continuation of pure or 'thorough' breeding from the Oriental imports, and it is unlikely that racing ability on the part of these purebreds was necessarily a consideration, since the Oriental type itself was unsuited to the long courses of the seventeenth and early eighteenth centuries. The decision of this group of breeders, located in Yorkshire and northern Lincolnshire, and perhaps centred on the stud of the d'Arcy family at Sedbergh, was contrary to the received wisdom of the seventeenth century, which believed that the Oriental or 'blood' horse would deteriorate in the English climate and therefore new ones must constantly be sought in their places of origin. To embark on the attempt to breed the Oriental type in England implied a strong belief that their quality was not just transmissible to their cross-bred offspring for one generation, but that the type would breed true, despite the change in its environment, for several generations, or indeed indefinitely. The strongest practical motive for the attempt was the huge expense attached to bringing in foreign horses, which in the case of Arabs and Turks remained difficult throughout the eighteenth century.[21]

The breed remained open to these Oriental and North African horses during the eighteenth century, although their use for siring Thoroughbreds fell sharply after 1750. The most obvious reason for this was that distinguished, race-proven, Oriental-type Thoroughbred stallions were by then available in large numbers and were used in preference to untried foreign horses because the Thoroughbred had now become the premier racehorse in its own right as the racing of young sprinters became increasingly the fashion. It had always been the Thoroughbred mares, foaled and raised at home in England, whose pedigree had been most important in the early days and the *General Stud Book*'s primary objective had been to trace the Thoroughbreds of the late eighteenth century to their original source mares.[22] The foundation mares upon which the early breeders worked could only have been part-bred, and in the first instance wholly English or mixed in origin. However, the breeding of replacement mares in closed herds in the seventeenth century probably meant that reasonable numbers of part-bred Barb and Arab mares were available to the foundation breeders of pure-bred Orientals in England when they began. The conscious attempt to pure-breed Orientals at home probably dates from after the Restoration since Newcastle gave no hint that he knew of any attempt to breed the Eastern type in England. Once a large population of Thoroughbreds had been evolved and home-bred sires preferred to imports, the pedigree of the sires became more important than that of the mares since, whatever the racing fraternity claimed about their attention to the mare, the traditional belief that the stallion was more important in stamping his properties on the offspring was certainly a dominant one among breeders for the turf.

The breeders of the Thoroughbred probably had some model of breed purity maintained among imported stock to give them some hint that breeding 'blood' horses in England would not lead to deterioration. The model was probably the continued breeding of fancy poultry and dogs from exotic locations in this country[23] – the sort of experience and observation which a working farmer concerned with agricultural livestock was less likely to have had. Farm livestock breeders in turn almost certainly used the Thoroughbred as a model upon which to base their own experiments. The use of the term 'blood' by the horse fanciers may itself be an indication of their belief in the hereditary principle since seed was ultimately supposed to derive from blood in some of the theories of generation current in the seventeenth century. They believed that 'blood' would transfer from the imports to native-cross offspring and then that 'blood' characters could be preserved against the vagaries of the climate and nutrition of northern Europe by careful preservation of the purity of the breed.[24] As this policy seemed to have been successful and properties were shown to descend from one generation to the next, the simple hereditary idea implicit in the term 'blood' led to a mystical belief that the lineage or descent of a horse governed its quality, although direct evidence might show that the animal concerned was a worthless hack. The original notion of the preservation of type had been unreasonably extended to embrace the idea that properties could descend intact in particular arrays over a long series of generations. Certainly by the mid-eighteenth century the worship of 'blood' by the examination of pedigree had often overcome common sense in the selection of horses and mares for mating.

In the foundation and creation of this 'thorough' bred 'blood' horse group, a degree of artificial and natural selection must have occurred since the animals were in a completely different climate and used for very different purposes from those for which they were originally bred in their homelands. The Thoroughbreds of the 1750s were therefore demonstrably different from the imported types from which they had been derived, if only because

Plate 6. A supposed portrait of the racehorse, *Eclipse*, painted in 1770 by George Stubbs. This is thought to be a study for the finished portrait of *Eclipse* at Newmarket with groom and jockey. Foaled on 1 April 1764 during the great eclipse in the Duke of Cumberland's stable at Windsor, *Eclipse* was one of the most famous eighteenth-century racehorses. He was in training for only two seasons in 1769 and 1770 and was said to have sired 334 winners at stud before he died in 1789. He belonged to the last generation of racehorses not to have been brought forward to race at two and three years old. See J. Egerton, *George Stubbs, 1724–1806* (Tate Gallery Publications, 1984), pp. 83, 133. Photograph from the Paul Mellon Collection, Upperville, Virginia.

they were blended from several different import races. These differences were sufficiently great by mid-century for the import of more Oriental genes to be counter productive. The local strain had been evolved and it is contended that very little radical change has occurred since.

If the Thoroughbred did not emerge as the dominant force in horse racing until the 1750s, racehorses before that date must have been of various types. These early eighteenth-century racers were presumably traditional half-bred English–Oriental horses and the remains of the older tradition of racing native riding horses, such as Galloways, Hobbies and Yorkshire saddle horses.[25] Some limited evidence can be produced to support this view.

The 6th Duke of Somerset, at Goodwood, appeared to be breeding successful racers in the traditional cross-bred fashion, paying no attention to pedigree and not aiming to produce 'thorough' bred animals. In 1701, Somerset did not own an Arab or Barb stallion of his own and used that of a neighbour, Sir William Goring of Burton. The only stallion that Somerset himself possessed was a horse called *Hautboy*, which may have been the same Newmarket racer which the Earl of Bristol wagered on in a match in April 1698, to his loss of £200. Somerset also owned a horse called *Young Yellow Jack*, possibly the same horse, or directly related, as that of the same name run by Lord Granby in 1704, when Bristol's horse, *Spider*, beat it in a match and won him £1200.[26] If these identifications are correct, it shows that Somerset's stud at this time was devoted to the production of top-class racers. But his stud mares were generally native and not distinguished by the possession of any genealogy. One of the mares was a *Foundered Yorkshire Mare*, another was referred to as the *Bamsteade Mare* and was probably the same as the *Banstead Down Mare* mentioned later in 1704, his own pad mare was bred from in that year, and in addition he possessed an animal called the *Gawden Filly*.[27] Only this *Gawden Filly* was covered by *Hautboy*, while the other mares were serviced by the neighbouring Arab. The latter series looks like the time-honoured cross of an Oriental stallion on to native mares to obtain cross-bred racers, while the doubt about the provenance of *Hautboy* and the *Gawden Filly* make that union hard to interpret.

A complete record of the matings at Petworth in 1704 has survived, a year when Somerset evidently possessed his own Arab and Turkish stallions.

Hautboy	x	*Banstead Down Mare*
		Large Flanders Mare
		Flanders Mare bought of Payne
Lame Grey	x	*Black Danish Mare*
Stoned Horse		*Bay Danish Mare*
Arabian	x	*Gawden Mare*
		Bay Bald Mare
		Lame Yorkshire Mare
		Chestnutt Hautboy Mare
		Flanders Mare that had the filly
Turk	x	*Burnett Galloway*
		North Country Mare
		Yellow Filly
		Pad Mare
		Brown Mare
		Hepshott Mare

All these matings were consistent with standard seventeenth-century practice. The draught coach mares have been mated with a blood racer, *Hautboy*, or the *Lame Grey* stallion, so that the racer itself has not been bred to obtain racers. These were apparently being sought from crosses of Oriental sires on native, unpedigreed mares, several of them from the regions where native saddle and road horses had long been bred. Such a procedure seems to have been adequate at this date for the generation of successful racehorses.[28]

At the end of the seventeenth and the beginning of the eighteenth century, one of the leading owners and breeders of racehorses in England was John Hervey, first Earl of Bristol, whose Diary and Expense Books reveal something of his racing activity between 1690 and 1714. In some of the earlier and smaller races Bristol rode his own horses, but already in the more expensive matches between horses weight was becoming significant and, although not excessively light, weights of between eight and ten stone were probably too low for Bristol to achieve. When he was riding himself he usually entered twelve-stone races. His establishment at Newmarket was already of the massive type associated with modern racing and the degree of Hervey's direct involvement there must have been very small. Some of the breeders, owners and horses mentioned by him found their way eventually into the *General Stud Book*, but a large number did not, suggesting that many contemporary racers were not of ancestral Thoroughbred type. Some of the horses of his own breeding or acquisition, which were reasonably cheap and undistinguished in their ancestry, seemed to do as well at Newmarket as the more expensive animals.

Later on in his career Hervey bought named mares at high prices and also a number of trained racers at very expensive rates from breeders who have subsequently gone down in the annals of the Thoroughbred, such as Leonard Childers and Anthony Leedes. Already the best 'thorough' bred stock had acquired a reputation and could only be obtained at prices comparable with those of stock imported directly from the Mediterranean source regions. In 1706 Bristol purchased three young horses from Leedes in Yorkshire. Significantly, while the pedigrees of few other of his purchases were noted, he recorded those of the animals he bought from Leedes very carefully. For a three-year-old filly, a two-year-old colt and a one-year-old colt he paid £150.[29] The Thoroughbred race soon afterwards produced the first of the legendary eighteenth-century racers in *Flying Childers*, foaled in 1714, and two generations later the group of famous foundation sires of the major breeding 'lines' which were foaled in the 1740s to 1760s.[30] These established the superiority of the breed on the racetrack under sprint conditions.

The group of late seventeenth-century breeders who were concerned in the first instance with the 'thorough' bred type were dealing with a very small number of imported horses. In a small, closed group, the problem or benefit of inbreeding would have presented itself at an early stage. Newcastle had advocated close inbreeding, but only in a closed herd of mares, with frequent outcrossing to new sires. The originators of the Thoroughbred were working with a small sample of both sires and dams and followed conventional prejudice in avoiding close relationships in their breeding, an additional reason for carefully recorded pedigrees. Certainly, d'Arcy petitioned to be allowed to import more Arab and Barb stallions so that the large influence of his own stock on all the surrounding stallions could be mitigated. He could not find sufficient local stallions of any note which were not of his own blood.[31]

Inbreeding of the type later popularised by Bakewell and actually practised by the Collings and Thomas Bates was certainly not applied in the origin of the Thoroughbred. A restricted group of foundation animals were employed, but they were genetically unrelated and apart from occasional close matings, the ancestry of many early Thoroughbreds showed a careful avoidance of incestuous breeding. Only in the late eighteenth century did inbreeding occur on any scale, notably in the experiments by Lord Derby at the time when it was the fashion among farm stock.[32] Even in nineteenth- and twentieth-century pedigrees, where common ancestors were frequent, they were distributed widely through time and if the inbreeding coefficients were correctly, rather than casually, assessed, the actual inbreeding indices would not be all that high. An analysis of all the mares held by Lord Godolphin over the period 1732 to 1754 shows that relatively little inbreeding had occurred, apart from one sharp and isolated incident of a parent–offspring cross.[33] Prior was probably right to suggest that the *General Stud Book* pedigree of the great *Flying Childers* was wrong and the more orthodox version given by Cuthbert Routh was to be preferred.[34] Another feature of the early Thoroughbred breeders that may be typical emerges from Godolphin's book: the practice of buying in mares at intervals with relatively few of the brood mares being home-bred, in contrast to seventeenth-century advice to select foundation mares and generate replacements from them over several generations.[35]

By the beginning of the eighteenth century considerable care and attention was lavished on the management of racehorses. Defoe visited Penkridge fair in Staffordshire, which had long had a reputation as the outlet for the best horses bred in Yorkshire and found:

that there were not less than an hundred jockeys and horse copers, as they call them there, from London, to buy horses for sale. Also an incredible number of gentlemen attended with their grooms to buy gallopers, or race horses, for their Newmarket sport.

The northern grooms were extraordinarily attentive in dressing and showing the horses:

they lie constantly in their stables with them, and feed them by weight and measure; keep them so clean and so fine, I mean in their bodies, as well as their outsides, that, in short, nothing can be more nice. There were several horses sold for 150 guineas a horse, but then they were such as were famous for the breed and known by their race, almost as well as the Arabians know the genealogy of their horses.[36]

On a visit to the centre of racehorse breeding country at Bedale in Yorkshire, Defoe remarked on the phenomenon of keeping pedigrees again and noted that the general Yorkshire gallopers were not yet entirely subservient to lengthy pedigrees:

the breeds of their horses, in this and the next county are so well known that though they do not preserve the pedigree of their horses for a succession of ages, as they say they do in Arabia and in Barbary, yet they christen their stallions here, and know them, and will advance the price of a horse according to the reputation of the horse he came of.[37]

The management of the brood mares was also carefully organised. Bradley mentioned that 'many' believed it better to house and feed a stallion carefully and bring mares to him for covering 'in hand' rather than allow the horse to run with the mares in a close. Such mares were themselves often kept for a month or two in stables on the same diet before mating, in an attempt to make conception more certain in the notoriously infertile matings of horse and mare. Furthermore 'some again are so careful' that they housed the pregnant mares from September onwards, feeding them on hay and mashes until foaling time, when mares and foals were allowed out to grass again. Clearly, the only people with the stabling, the staff and the finance to attend to their mares on this individual basis were those breeding racers.[38] Hale confirmed in mid-century that this was indeed the practice of the breeders of high-class horses.[39]

If the precise methods of the breeders of Thoroughbreds and other racers at the beginning of the century remain obscure, rather more can be discovered about breeders in mid-century from the works of William Osmer and Richard Wall. Wall published his *Dissertation on the Breeding of Horses upon Philosophical and Experimental Principles* in about 1758, while Osmer produced the first edition of his *A Dissertation on Horses* in 1756, in which he propounded novel views about racehorse breeding. A second edition was produced in 1757 and he published a separate dissertation and reply to his critics as an appendix to his *Treatise on the Diseases and Lameness of Horses*, in 1761. Both authors took contemporary breeders to task for the inadequacy of their methods, but Osmer attracted more obloquy because he was anxious to discount altogether the significance of heredity or 'blood' in the excellence of a racehorse, something with which the jockeys, now used to the pedigreed Thoroughbred, could not possibly agree.

Wall started by noting that the number of racehorses which could be called top-class was a very small proportion of those bred, and he felt that his

Plate 7. Otho, with John Larkin up, painted in 1768 by George Stubbs. This apparently accurate painting by Stubbs may be taken to represent the form of a good average racehorse in the mid-eighteenth century. Horse and rider are shown by a rubbing-down house at Newmarket which features in several other Stubbs racehorse portraits. See Egerton, *George Stubbs,* pp. 88–9. Photograph from the Tate Gallery, London.

own system would increase the fraction of very fast animals in any generation. He believed the problem lay with methods, typical of many breeders, in which chance was allowed to play too great a part. He thought that many of them who pursued methods of line breeding based on pedigree, the type of method used in Thoroughbred breeding throughout the nineteenth and much of the twentieth centuries and which the breeders thought of as 'nicking qualities in the cross', were merely disguising the chance nature of their methods.[40] Wall optimistically supposed that horse breeding could be made more successful by the application of logical methods which excluded chance. This betrayed his lack of practical breeding experience, whereas the farrier, Osmer, was far more prepared to accept the inevitable presence of random variation in breeding plans.

Wall listed a series of methods which he described as standard but which he felt were erroneous or illogical.

(a) Some breeders had a fondness for a family of a particular horse or a partiality to a particular blood. Here, the pedigree of the stallion alone was considered and no thought was given to its physiological properties, nor was any concern at all shown for the mares.

(b) Larger studs were often free of that particular fault because they had mares and horses of several genealogies. However, very few of the mares in large studs had ever been trained and consequently their racing potential was completely unknown. Such breeders would pay considerable attention to the pedigree and performance of the sire but would merely select their mares on some abstract basis for 'nicking' with the good qualities of the stallion, whereas it would be far better if the racing quality of the mare were known.

(c) Some breeders put all their mares to a particular stallion that had acquired a good reputation as a racer, while still giving no thought to the mare. They believed that a good racer could result provided the mare was herself a Thoroughbred.

(d) Some young breeders used half-bred mares with a Thoroughbred stallion, hoping there would be sufficient virtue provided by the blood horse to generate a good racer.[41]

One may conclude, reading between the lines of Wall's critique, that the majority of racing stallions were now themselves racers of Thoroughbred type and that foreign stallions were seldom employed. Conversely, Thoroughbreds did not have a monopoly since young hopefuls could still enter part-breds for races and hope to find themselves with winners.

The major plank of Wall's criticism was the lack of attention paid to the mare compared with the stallion. He believed that a foal inherited its properties equally from both its parents, and these parents themselves had quality by virtue of the ancestors in their pedigree, so that the performance, symmetry, proportion, and pedigree of both sire and dam should be examined carefully. Only animals which showed excellence in all categories should be bred from. They should be matched so that their faults and good points in all departments 'nicked' or cancelled each other out, rather than reinforcing each other. This last was conventional wisdom among most breeders. Hereditary defects and all matings between related or sibling animals were to be studiously avoided.[42]

Even the best and most rational choice of parents could still lead to very poor offspring. In many cases Wall ascribed this failure to poor management of the dam or the foal so that its full potential was not realised. He was also forced to concede that even if the management was of the same standard as the selection, his predicted result of an offspring intermediate between the two parents would sometimes not follow. Pure chance could lead to a poor foal, or indeed to a much better one than the quality of the parents warranted.[43] The expectation of median quality meant that nothing of any note could be expected except from the mating of the very best animals. His specific example of the production of median quality was the mating of a slow country mare with a Barb, where the offspring would be expected to perform at an average of the speed of the two parents.[44]

Wall then said many breeders regarded their mares as largely responsible for the quality of their offspring, whereas he had been berating them for neglecting the mares only a few pages before. However, he may have been right in the sense that most owners were biased in favour of their own material even if there was no rational basis for their opinion. Most of them would argue that this influence was the result of the experience of the foal in the womb and its suckling of the dam's milk. Wall believed this maternal role could have no influence on the foal since he believed hereditary characters were fixed at conception.[45]

Wall's firm belief in the importance of heredity in equine quality meant that he did not think horses bred in Arabia possessed characters as a mere consequence of living in that country and suggested there were large numbers of worthless animals there and that some of the declining value of Oriental imports at that date was due to the poor selection of animals in their home regions. Likewise, he claimed that racehorses every bit as good as those of Yorkshire had been bred in other parts of the country and that there was nothing purely environmental in the superiority of Yorkshire as a horse-breeding county.[46]

In contrast to Wall's acceptance of the virtue and reality of heredity or 'blood', Osmer purported to believe that the virtues and character of horses were entirely independent of this non-material transmission. Seeing that the neglect of conformation and performance was careless and unsound practice as more and more breeders tried to find a way round the lottery of genetic reassortment by concentrating on pedigree, he rejected any hereditary component whatsoever in the production of performance in the horse. Instead, he was driven to rely on the view that properties were dependent upon climate, nutrition and geography,[47] together with his own opinion that a certain set of physical characters in the conformation of the animal were significant for the production of speed. This view of conformation was not original, but Osmer's judgement of what these conformational features were differed from the conventional opinion, which remained based on aesthetics, rather than being a rational assessment of form in relation to function.

Osmer expressed this view by saying that the virtue of a horse was altogether mechanical and not in the 'blood'. Mechanical form meant not just the proportion and symmetry of the whole but the formation of the limbs and joints and the setting-on of the limbs to the body, which in Osmer's view should be of such a form and position that many Thoroughbred breeders would regard the resultant animal as grossly at fault and downright ugly. Osmer believed that such ugly and ill-shaped and even ill-going animals were good by virtue of their form, while the conventional wisdom was that they were good despite their form because they were of the correct racing family and 'blood', so that their virtue was inherited independently of their mechanical shape. In searching for a model of conformation for speed, Osmer was much impressed by greyhounds and hares and argued that the nearer the horse approached to the shape of a greyhound, the faster he would be. He believed the virtue of the Arab as a breeder of running horses resided in his form and he attacked the adherents of abstract inheritance of 'blood' with the excellent point that full siblings in both animals and humans who shared identical 'blood' often turned out to be physically dissimilar and to have variable performance as runners as a result. He was in something of a dilemma himself in his attempts to explain the origin of such differences, being forced to suggest that it was caused by damage to the womb of the dam, or the seed of the horse.[48]

Osmer claimed he would rather breed from any horse with the correct conformation as he saw it, but with no thought to its genealogy, than from a high-bred animal whose conformation was entirely wrong. He believed the explanation of the fact that a non-racer from a good family sometimes produced a good racer lay in the idea of his mechanical conformation being corrected by what was inherited from the mare.[49] He believed random breeding would have given just as good results as the careful organisation of lineage crosses on the basis of pedigree analysis, in which Thoroughbred breeders were then beginning to indulge:

Is it not the truth ... that the very best reputed breed horses and mares in the kingdom cannot run at all? Yet still they serve to breed from for the sake of the blood or the cross   hence it is that the breed of horses in this kingdom, is little superior to a parcel of hackneys, in comparison with what they might be, if well understood.[50]

He criticised even more cogently the designation of some families as running horses and others as not, for this was usually based on some arbitrary decision in relation to one key animal in the genealogy. Even where the reputation of this one animal was justified by its performance, there could be no question of the inheritance of its properties through an infinity of generations.[51]

Perhaps the quotation which best sums up Osmer's position would be the following:

So there is nothing in blood – no – nothing at all, independent of form and matter, as the sportsmen say there is – But the Arabian horses, being better constituted for action than other horses, do by means hereof excel all others, and each also according to the degrees of difference in their form and constituent parts.[52]

To suggest that sportsmen believed that blood operated independently of form was a distortion and oversimplification of their position, but it was certainly true that Osmer chose to attack a general belief in the power of the pedigree, which was to grow more powerful as the years went on, and he was correct to point out the general worthlessness of much which was to be done in the name of rational breeding on the principles of genealogy.

Work horses with style: carriage and draught horses in the eighteenth century

The majority of coach and waggon horses for the large London market were supplied from the Midland grazing counties of Staffordshire, Derbyshire, Nottinghamshire and Leicestershire. The best working colts and fillies bred in these regions were transferred to London by the elaboration of a trade which had originated in the seventeenth century.[53] The horses were sold in the North Midland fairs to farmers from the Central and South Midland arable regions, who had not the pasture or the time to breed for themselves. These animals were worked for some two or three years and then disposed of to London for coach and waggon work. Their labour paid for their keep on the farm and they were sold at the end of this period at a profit. Mortimer described how Hertfordshire farmers bought up colts from the Leicestershire horse fairs for this purpose.[54] Laurence stated that in Northamptonshire, two- or three-year-old colts were purchased from Leicestershire and Nottinghamshire, taught to work in harness and worked lightly for a year or two, before being resold in Northamptonshire fairs for the London coach trade at prices between £20 and £40. The farmers could make a good profit from this provided they were reasonable judges of horse flesh and did not work their animals too hard. Some drudges were always kept on the farm for the really tough work.[55] Bradley added that the most skilful Midland farmers took care only to buy three-year-olds, since two-year-olds were thought to suffer from the change of air and food so early in their development. He also commented that some of the Northamptonshire farmers were now breeding Leicestershire Black horses themselves, rather than merely buying them in from that county.[56]

These Leicestershire breeders were the leaders in the development of the Midland Black cart breed, which had been heavily influenced by Flemish imports from the sixteenth century onwards.[57] Yearlings were sold through autumn fairs at Burton, Rugby and Ashburn and then sold on at two-and-a-half years at Stafford and Rugby fairs to the South Midland and western arable zones, where they were broken, worked and sold on to London when nearly mature at five or six years old for waggons, coaches and cavalry remounts.[58] The hint that some Northamptonshire farmers too were breeding the Midland Black in the 1720s had become a reality by 1770 when Arthur Young noticed that the breeding of these animals around Naseby was a considerable business and two-year-old colts were disposed of at Harborough fair for prices from £10 or £12 upwards.[59] However, in the 1790s, most of the county was still engaged in the classic trade of buying horses in from Derbyshire, Lincolnshire and Yorkshire at two or three years old, working them, and then selling them on to London. When five or six years old, those good enough for the coach trade would fetch £40 (rather more than in Laurence's day) while the inferior or grosser animals could be sold for the dray or waggon at £30 and as cavalry horses for about £25.[60] The huge Midland Blacks were also bred extensively in the Lincolnshire fens for sale into the South Midlands, at £10 to £12 for yearlings and £18 to £20 as rising two-year-olds.[61]

The position of Yorkshire in the supply of this lucrative trade is unclear. Worledge had said in the seventeenth century that Yorkshire was a main source of middle-sized draught horses for London coach work, but none of the early eighteenth-century authors mentioned Yorkshire in this context. Marshall implied in the 1780s that the main fame of Yorkshire had always been its saddle horses and hunters.[62] The location of so much Thoroughbred breeding in the county supports this contention, since the rootstock of native racers had been fast riding horses. In the Vale of Pickering, from about 1750, classic Yorkshire work horses had been lightened and converted into hunters by an over-enthusiastic infusion of Thoroughbred genes, notably from the stallion, *Jalap*, a grandson of the *Godolphin Arabian*.[63] By the 1780s the native saddle horses were being beefed up again to create animals suitable for fashionable light carriage work. At this date Yorkshire horses were being sent straight to London, either overland in strings or shipped out by sea through the port of Hull, and were not going through the period of farm work in the Midlands, which suggests Yorkshire had not been a supplier of coach and waggon horses earlier in the century.[64]

The curse of overfineness and lightness introduced by too much Thoroughbred crossing was still making Yorkshire horses too light for farm work in the 1790s,[65] although the presence of 'blood' genes certainly improved their value as fashionable carriage horses. Coach horses in the late eighteenth century were increasingly supplied from North Yorkshire by the Cleveland Bay, where less Thoroughbred crossing had occurred than elsewhere, while the fashion for an even lighter framed carriage animal was provided by the Yorkshire Coach Horse, an F_1 cross of the Cleveland Bay with the Thoroughbred. These breeds displaced the Midland Black from the coach and confined it to the waggon, so that whereas early in the century coaches were generally pulled by heavy, black horses, by the end of the century the carriages of the gentry were pulled by light, bay horses.[66]

Plate 8. Mares and foals in a river landscape, painted between 1763 and 1768 by George Stubbs. The horses portrayed are taken from an earlier commission for Lord Rockingham of a frieze of mares and foals. Stubbs has edited out some animals from the previous composition and added an idealised pastoral landscape to create a slightly chocolate-box idyll of brood mares in a gentleman's park. See Egerton, *George Stubbs*, pp. 126–7. Photograph from the Tate Gallery, London.

Apart from the Yorkshire Bays sold direct to London, Lincolnshire graziers also bought some of them as colts, pastured them for a year and sold them on to London, having first docked and nicked them. From Lincolnshire, the London dealers bought them at £35 to £40 each and sold them after a month's training in a break or carriage for gentleman's coaches at 70 to 80 guineas each. Alternatively, the Yorkshire breeders retained them until they were four years old and themselves docked, nicked and fattened the animals on bran mashes for three months. They then drew the corner juvenile teeth of the four-year-olds so that the adult teeth denoting a five-year-old would appear early. These animals were sold into carriages as nearly mature five-year-olds. The London dealers insisted that the Yorkshire and Lincolnshire keepers did this and the breakdown of immature horses under the strain of coaching on hard pavements was apparently considerable, thus increasing the demand for horses.

Regions beyond the Midland counties which lay between the Midland Black and Yorkshire Bay breeding grounds and London, had to breed their own work horses. The most famous of these in the eighteenth century was the Suffolk Sorrel breed. According to Young, this horse had been a singularly ugly animal in the middle of the century but well suited to its function as a farm work horse. According to Marshall, it still was a very ugly animal in the 1780s, being no match for the aesthetics of the Midland Black, even though this latter breed was one to which Marshall was little inclined. In the 1790s Young said its shape had been improved in an attempt to render it suitable for the coach trade, so that a good five- or six-year-old gelding could sell for 30 or 40 guineas, although as a pure draught animal it had probably suffered from the lightening of its frame. The old breed had been barrel-shaped, short-bodied and legged and higher on the rump than the shoulders.

Suffolk Sorrels were the subject of a considerable fancy among local farmers, both in their breeding and in the assembly of teams for drawing matches, where the whole team was trained to a very high degree of obedience and control. They were undoubtedly very strong for their weight and better general farm and plough horses than the Midland Blacks, which had become too heavy for this purpose by the second half of the century.[67] In the neighbouring county of Norfolk, the Suffolk Sorrels were being crossed into the local mixed breed of the late eighteenth century, the Midland Black having spread down into Norfolk and hybridised with the local, small draught horse. Marshall was pleased to see the introduction of the Suffolk as a fashionable top cross in Norfolk and an antidote to the unnecessary size and ranginess of the Black breed.[68]

The Suffolk Sorrels would eventually give rise to the modern Suffolk Punch. The ancestors of the third modern heavy breed were already gaining a reputation as well. The Clydesdale horses of southern Scotland were being imported into Northumberland for draught work in the 1790s, at which time they were animals averaging $15\frac{1}{2}$ to 16 hands high. The local breed was not so distinguished, being described as black, heavy and rough legged.[69] Almost every county still possessed vestiges of its local breed, although few of these impinged upon the pens of contemporary commentators.[70] There was only one other area where draught horses of more than local significance seem to have been raised and that was in Somerset and Gloucestershire. Colts were bred in large numbers on the Somerset Moors, sold into the northern counties and bought up by Midland dealers for distribution to London, as an offshoot of the classic two-stage Midland trade. The change in fashion from Black to Bay horses had evidently finished this West Country trade, so that when Marshall visited Gloucestershire in the 1780s, the farm horses were all home-bred Blacks in the Vale of Gloucester, with six-year-old cart horses only selling for local work at between £25 and £30.[71]

At least one type of light coach horse (the Yorkshire) was derived from crossing Thoroughbreds onto a heavier draught type. This sort of cross, from the mid-eighteenth century onwards, became the major system for the generation of saddle horses and hacks for hunting, riding and cavalry purposes, replacing, at least among the gentry and the better-off farmers, the use of light work breeds for this function. The enthusiasm with which these cross-breds were adopted suggests that they filled a gap in the market, perhaps not fully satisfied since the better breeders had abandoned the production of ambling and pacing horses for riding in the seventeenth century. Most of the lighter local breeds of work horse made adequate, if not very dashing, riding mounts. John Ball evidently regarded such an extravagance as owning a half-bred saddle horse in a very poor light. As far as he was concerned, any reject carriage horse of a light frame would do.[72] However, the concept of using 'blood' and later Thoroughbred horses to obtain spirited half-breds, was a characteristic of professional and larger farming people even at the end of the seventeenth century. By the end of the eighteenth century the use of Thoroughbred stallions to get half-bred hunters was commonplace.[73]

Thoroughbred horses were also used to create a new type of specialist road horse in the late 1700s for fast riding or the pulling of light carriages at speed: the Trotter. This type of animal was to become very popular in America in the nineteenth century. Most trotting breeds owed their origin to Thoroughbred crosses onto local mares, and among the earliest areas to specialise in this form were South Lincolnshire and the surrounding parts of Huntingdonshire, Cambridgeshire and western Norfolk, probably using the Midland Black mare as a basis on the female side. By the 1790s wagered trotting matches were starting in the Huntingdon/Cambridge region. Trotting matches with teams of light draught horses were also common in the rest of Norfolk.[74]

In contrast to the seventeenth century, when little advice was offered to ordinary farmers about the breeding of horses, the textbook writers of the eighteenth century directed a considerable amount of their attention in this direction. One reason for this was that horses on farms had largely replaced oxen as a source of power by this date, and far more farmers kept a stable of horses than had been the case in the seventeenth century. Another was the increased use of coaches, which produced so large a demand that the source had to be ordinary farm stock rather than the studs of high-class breeders.

Most of these writers could not resist repeating the advice of Blundeville and the Italian school on the suitable crosses of exotic horses to produce animals for various purposes, although most of them spent little time on this irrelevant subject and advised their agricultural readers to stick to native English horses. Implicit in some of their discussion was the age-old problem of inheritance versus environment in the control of the nature of the horse. On the whole most seem to have accepted that the real determinants were the parent animals, not the place or conditions under which they bred. Only

Ellis felt compelled to state that the pre-eminence of the northern counties in the production of horses was partly caused by the climate and nutrition in these regions, although he was also aware that rents and careful breeding also had a great deal to do with it.[75] The general advice given by the didactic authors might be summarised as follows.

The foal would resemble its parents, so selection of these was important. On the whole it was thought that the mare had most to contribute to size and form and that especial care should therefore be taken to have large-bellied and well-formed mares. The antique Aristotelian corollary that the stallion had more to do with the character and personality was generally not stated, although it was felt that both parents should have a lively and brisk nature so that the foal would inherit these properties as well. Both parents should be healthy and sound so that no infirmities were passed on.[76] On the question of the age of mares and stallions for breeding, most authors thought that both parents should be full-sized and mature before starting and that once past their prime they should not be bred from. They agreed that the breeding life-span of both mares and stallions would be greatly prolonged if neither were used for work.[77] But most brood mares and stoned horses on farms were expected to work as well as breed for their living.[78]

Many authors stressed that only sound horses should be bred from, although it appeared in practice that all too often the animals used were those no longer fit for other purposes. In the later years of the century, the horse coursers and jockeys who paraded Thoroughbreds round the country for the use of farmers, generally used cast-offs, blood vessel breakers, lame or otherwise defective animals for which the racing fraternity had no further use.[79] These aged hacks were sometimes merely accidentally damaged animals, but they often carried defects and deficiencies which were hereditary. There were frequent references to the pernicious habit of using defective breedstock among ordinary farm horses. The use of blind stallions and mares seems to have been particularly prevalent. Laurence first pointed out that coach horses, bred from country animals, were too often liable to fall blind in one or both eyes at four or five years of age.[80] While some attributed this to overfeeding with 'hot' foods such as peas, others saw its origin in the use of blind mares and stallions for breeding, which was already common by this date. The blindness was almost certainly hereditary. There was also a belief, first articulated by Bradley, that the foals of old stallions were likely to become blind.[81] Ellis elaborated this point and extended it to include the produce of overworked stallions since, following Bradley, he believed the overserving of mares (by which he meant more than 16 in a season) led to the same weakness as that of old age. He also thought that certain coat colours were correlated with the generation of blind foals.[82] It was claimed that by the middle of the century those who bred foals for the coach market

Plate 9. The Duke of Ancaster's bay stallion *Blank* walking towards a mare, painted *c.* 1770 by William Shaw (*fl.* 1758–72). The stallion has the curious name because Ancaster refused to register the names of any of his stud horses. The horse is being led by a groom towards a mare in the background. The mating will be carefully supervised and recorded. Shaw's portrait is stylised and wooden compared with contemporary horse portraits by Stubbs. In at least one other similar subject, Shaw used an identical pose and form for the horse, but viewed from the other side, with the coat colour changed. See Egerton, *British sporting and animal paintings*, pp. 112–13.
          Photograph from the Paul Mellon Collection, Upperville, Virginia.

were deliberately choosing blind stallions and that a number of these were always in circulation in the markets of Leicestershire, Northamptonshire and Staffordshire during the season. The blindness did not usually appear until the horses were five or more years old, so that when they left the dealers they were perfectly sound. The buyers never suspected the source of the blindness and the breeders and dealers had a vested interest in continuing the practice, in order to inflate the number of animals required.[83] That blindness in workhorses was a problem can be confirmed from contemporary farm manuscripts. For instance, when two small farms reverted to Henry Hunter of Swallowfield in Berkshire in 1776, of nine horses on the farm (admittedly a pretty decrepit group) there was one blind four-year-old horse, another blind ten-year-old, an eight-year-old blind mare and another old mare blind in one eye.[84]

In the management of brood mares most of the advice given was conventional and derived from earlier sources. The ancient technique of using a teaser stallion to assess the mating state of the mare was often repeated, as was the classical advice to throw cold water over the mare after horsing to ensure the seed 'took'. Little reference was now made to the phases of the moon and none at all to left- and right-sidedness in sex determination. In practice, far more attention seems to have been paid to the stallion in the breeding of the average work horse, especially when Thoroughbred stallions became available for hire at reasonable rates among local farmers and breeders from mid-century on.[85] The habit of concern about the pedigree of the stallion spread from the jockeys to the ordinary farmer and by the late 1700s the pedigrees of coach and cart stallions began to be as carefully considered as those of blood horses. The show and sale practices of the Thoroughbred breeders had also spread to the breeders of draught horses by the end of the century.[86] One feature of horse breeding which had originated in the seventeenth century, the payment of a fee for the hire of a stallion to service mares, expanded considerably in the eighteenth century. The point of this practice for the owner of a couple of mares was obvious, but evidently the use of 'foreign' stallions was also widespread among those who owned sufficient mares to justify the upkeep of a stallion. There are three possible reasons for the beginning of this procedure: the desire to use an animal of a quality superior to one's own mares, to follow the precepts of the Italian school in the crossing of exotic stallions on to local mares, or merely to ensure the health of a race of mares was not impaired by the use of a stallion of one's own breeding. Whatever the reasons for the origin of stallion hiring, it is evident that some owners specialised in maintaining and hiring stallions, while others concentrated on holding brood mares from which to generate foals. A breeder who maintained a herd of brood mares on grass as a specialist occupation probably had little use for a stallion apart from his service and may have found it cheaper and less troublesome to hire a stallion for a fee during the breeding season. Others, especially those whose interests were in arable farming or whose pasture was more profitably employed in pursuits other than horse breeding, could keep a stallion among their work horses to earn money from hire fees during the season. The hiring of stallions acquired another level of complexity during the eighteenth century with the hire of animals to regions remote from those where they were normally kept.

Local stallion breeders or casual stallion owners rented their animals out for the breeding season to travelling horse copers, coursers or jockeys, who then travelled the horses to distant regions, showing them at fairs or advertising them in the local press. They charged covering fees which would meet the costs of their original hire, their travel and produce a profit as well. At the end of the season they returned the horses to their owners and presumably operated for the rest of the year as ordinary horse dealers. Marshall thought this system had started on a local scale at the beginning of the century to save a farmer with a likely-looking horse having to take it round the fairs to obtain lettings, when he should have been employed elsewhere on the general husbandry of the farm. He also believed the model of the travelling horse jockey with his stallion for hire to have been the origin of the trade in ram and bull letting in the Midland counties and Lincolnshire.[87] Evidence for the use of hired stallions is plentiful from contemporary eighteenth-century sources.

In the late seventeenth century the standard fee for the 'leap of a stallion' seems to have been about 2s 6d, with the service of a high-class animal running at 10s to 15s.[88] The stallions appear to have been local and the mares to have travelled to the horse for cover. Nicholas Blundell, in the early eighteenth century, owned a small group of mares from which he bred foals, as well as a team of coach horses. In his will he left his breeding mares, described as coach mares and colts, to his son,[89] so it is possible the brood mares described in his Diurnall were coach mares. In 1712 only one mare, *Bess*, was covered successfully – by a Bay of Thomas Blaxter's – but the foal was lost in November. Thomas Bigarstaff, from Liverpool, brought his stallion up to Little Crosby and tried three other mares of Blundell's in June, but they were not in oestrus. In 1713 Blundell was more successful, this time taking the mares down to Bigarstaff's Bay stallion six miles away in Liverpool in April and May. All three mares foaled successfully in 1714. That year attempts were made again to cover the mares with Bigarstaff's horse, both in Liverpool and at Little Crosby, where several of Blundell's neighbours' mares were tried as well. Blundell paid him 2s 6d per foal for the three produced in the previous season, so the standard covering fee had not changed from the seventeenth century and the payment did not fall due until the foals had been born. Bigarstaff was evidently not merely renting the service of his stallion to Blundell as a friend but was providing a commercial service for small farmers and gentlemen in and around Liverpool.

In the 1720s Blundell's mares were horsed by stallions based at Ormskirk, some 8 miles to the north-east. In 1726 two of his mares were covered by a Black stallion standing there. This horse was owned or travelled by one Thurston Heskins and the animal was resident at Whittle-le-Woods, near Preston, 15 miles further north-east of Ormskirk. The following year he used the same stallion again, and also a Bay called *Jack*, which was said to have been got by *Ruffler* out of a mare of a Mr Win. It sounds as if Blundell was tempted to use a stallion of racing stock and was widening his horizons geographically and in terms of the class of stallion that he was prepared to use. As early as 1715 he had had a leap of a blood horse owned by Lord Molyneux, who trained his horses locally at Little Crosby. The horse was called *D'Arcy* and therefore probably a blood horse or ancestral Thorough-bred. In the 1720s Blundell was prepared to travel as far as Ashbourne fair in the Derbyshire Peak to buy coach horses and Galloways, paying 17 guineas for a coach horse there in 1726. From Little Crosby to Ashbourne was well over 100 miles. Blundell appears to have been a perfect example of the small gentleman with a taste for coach horses and hunters, and the grounds and time to indulge in the hobby of horse breeding.[90]

While Blundell was branching out and engaging the services of an occasional blood horse, the owners of the Camer Estate at Meopham in Kent were breeding, in 1718, from two mares aged seven and eight. In the following few years these mares were sometimes horsed by local stallions, sometimes by those owned by relatives and sometimes by their own stoned horse. However, in 1721, the owner of the farm was prepared to pay a guinea for the service of a horse belonging to Mr George Polley, a price at this date which suggests the horse must have been a 'blood' animal.[91] In the same county in 1739 a fee of 11s 6d was paid for the use of an unspecified stallion.[92] In the 1740s advertisements for travelling blood horses began to appear in the local Kent press at a standard fee of half a guinea and 1s for the groom, the same fee that the Faunce Estate had paid in 1739, which suggests that that horse too may have been of racing extraction. Whether the horse called *Tinker*, advertised in the hands of Isaac Kemp, near Sittingbourne, as a fine grey who had won several King's Plates was really a blood horse of any quality is another matter, but it seems that half a guinea was settling down as a standard fee for a good local or travelling stallion by the mid-eighteenth century.[93]

At Coton Hall in Shropshire in 1745 one of the mares was covered at the rate of 10s 6d, while in the following year the fee for the production of a foal from a mare was to be one guinea, but nothing was to be paid if there was no result, a practice which seemed to be dropped in most places in favour of a reduced fee if the mare was barren. In 1760 the Coton Hall fancy was somewhat more expensive. They sent a mare called *Strawberry* to a Mr Tattershall at a fee of two guineas to be covered by the horse *Young Traveller*. Their mare was boarded out at 2s 6d a week and there were other payments to servants for taking her to and from the horse. In 1763 another mare was covered at three guineas by a horse called *Genius*, presumably another blood horse, while a local stallion commanded only 6s 6d a leap.[94] At Waldershare in Kent another local stallion covered for 5s in 1753,[95] while in the Weald the Pattenson family paid a similar fee of 6s for the use of a local horse in 1767. In 1768 their brood mare was put to a local two-yearling colt, possibly got by a Thoroughbred, since the fee charged was one guinea for a successful foaling. In 1770, on the same farm, a saddle mare was put to a travelling stallion from Yorkshire for a fee of 17s 6d and a Sorrel mare to

Plate 10. Bay Ascham, a stallion, led through a gate to a mare, painted between 1802 and 1804 by J. L. Agasse (1767–1849). After settling in England permanently in 1800, the Swiss artist, Agasse, had considerable success painting in the manner of Stubbs. Although this picture lies outside the period covered by this book, it provides a far more realistic view of the preliminaries to mating than Shaw's picture (Plate 9). The stallion, a Thoroughbred, although never trained for racing, has the poise, with his neck arched and the curled lip of an animal which has caught scent of the mare. See Egerton, *British sporting and animal paintings*, pp. 181–2. Photograph from the Paul Mellon Collection, Upperville, Virginia.

another Yorkshireman for 10s 6d.[96] The reputation of Yorkshire for horses and its position in the centre of Thoroughbred development meant that it was worth Yorkshire jockeys travelling as far as Kent with their charges by this date. These horses were probably inferior blood animals, to judge from their prices. In Yorkshire itself, in 1773 at Leeds, an imported six-year-old Arab Chestnut was charged at five guineas a mare and 2s 6d for the servant.[97]

All the evidence so far presented has come from the users of stallions and not from the owners or jockeys. William Tompson of Forge Farm, Abbot's Bromley, in Staffordshire, was in the central region for the rearing of cart and coach horses, but worked a mixed and not a pastoral holding. He therefore specialised in hiring out horses to local mares – in 1758 and several subsequent years at the rate of 10s 6d if the service was successful, or 5s 6d if no foal resulted. In 1761 the terms were almost the same at 10s 6d for a foal and 5s 3d barren, but by 1762 the rates had fallen to 8s and 4s respectively. The horses concerned covered from April to early July in most years. In 1758 a total of 100 mares were covered, 33 of them requiring repeat services, but no hint was given in the cash book as to how many stallions Tompson owned, but it was probably at least five or six. If a 75% success rate for horsings is assumed, the income from this source in 1758 would have amounted to approximately £50.[98] In Derbyshire, Sir Robert Burdett was hiring out stoned horses as a way of recovering some of the cost of running his stable. Burdett himself sometimes ran to having his own mares covered by blood horses, paying £5 4s for the privilege in 1755. It is probable that his other mares were covered by his own single stallion, which he also hired out to others at the rate of half a guinea and occasionally a guinea a time. In most years the income from this source offset between a fifth and a quarter of the costs of running the stable. The average annual income was £10.44, equivalent to the stallion covering about twenty mares in additon to his work in servicing Sir Robert's own mares. As a stallion, in contemporary terms he seems to have been fairly heavily worked.[99]

Literary sources slightly later in the century gave figures for the price of a stallion to leap farm mares. Lord Kames estimated the expense at 15s a leap in 1766, while a correspondent to the *Scots Farmer* in 1773 worked on the traditional basis of 10s 6d a leap, or about a guinea a live foal, assuming one foal per two mares per year.[100] General fees for the use of local and travelling stallions certainly remained in this general region in the last thirty years of the century. They ranged from 7s 6d to £1 3s 6d in Kent and Sussex.[101] In the Midland Black breeding regions of Northamptonshire in the late 1760s, farmers would go as high as two guineas to have their mares covered by the best stallions of the breed.[102] At this time Robert Bakewell was letting stallions of the same breed at up to five guineas a mare.[103] In line with the fashion for Dishley stock in other areas, Bakewell's stallions were subsequently available only at great expense.[104] These other areas, of course, were the agricultural species of sheep and cow. He provides a form of link to allow us to move from the sporting animals of the gentry to the productive livestock of the farmer. The next chapters will give some consideration to the breeding of the farmer's beasts.

6

Cattle breeding: dairymen, graziers and the techniques of their 'fancy'

The cattle trade and the distribution of breeds

During the Middle Ages the primary purpose of keeping cattle was for draught. The object was to produce oxen for the cart and plough. The generation of replacement animals provided excess milk as a by-product, some portion of which may have been consumed fresh, but most of it was usually converted into butter and cheese. The only source of beef was from culled-out or aged animals of no further use for either work or breeding. Some hint of a regional stratification of the cattle trade was provided by the presence of specialised cattle-rearing enterprises in the North-West, whose function may have been to supply replacement plough animals to the Midland counties.[1]

During the sixteenth and seventeenth centuries this stratification became more and more pronounced. Replacement cattle for all purposes were increasingly derived from the pastoral regions in the North and West of Britain and used to stock the wastes and pastures of the predominantly arable southern and eastern lowlands.[2] An inevitable by-product of the growing export of animals from the West and the North was an increase in the volume of milk produced. A new market for dairy products encouraged regional specialisation in dairying, first in the peripheral rearing counties themselves and then in the counties nearby where grass growth and agrarian conditions were suitable for cattle feeding. Prominent among such dairy regions were the Lancashire Plain and Cheshire in the North-West; Somerset, Gloucestershire and North Wiltshire in the West Country; Suffolk; the grassed down areas of the Midland counties, notably Staffordshire and Derbyshire; and some specialised patches of Cambridgeshire.[3] In Essex and Hertfordshire, milk was not taken from the cow, but the calf left with its dam and killed as veal for the London market. At the same time the major market for store animals was changing from the supply of replacement plough beasts to stores for fatting, mainly on grass, but to a lesser extent in stalls or on other foodstuffs.[4] The trade of grazier became an increasingly profitable occupation and the rise of its importance was marked by the passage of Acts of Parliament[5] and by the description of grazing skills in the contemporary literature. As early as the 1530s Fitzherbert stressed that if lean cattle were purchased for feeding, they would fat better the younger they were, suggesting a beef market made up exclusively of old and worn-out stock was already a thing of the past. More significantly, he suggested sensible features to look and feel for in lean beasts which would be indicative of fatting quality.[6] Harrison made the same point about feeling a beast in order to judge the quality of the carcass in the 1580s.

If they doo but see an ox or bullocke, and come to the feeling of him, they will guesse at his weight, and how manye score or stone of flesh and tallow he beareth, how the butcher may live by the sale, and what he may have for the skin and tallow; which is a point of skill not commonly practised heretofore.[7]

The graziers were evidently anxious to develop some rule of thumb for judging the ratio of tallow to meat in the prospective carcass of the live animal in order to arrive at a better estimate of their value, for profit came from the sale of both these commodities as there was a large demand for animal fats and oils for lubrication. The response by the graziers to the question of the type of animal to aim for settled firmly in favour of large size, rather than rapid fatting or good food conversion. Thus, Markham advised dairy herdsmen to have large animals so that, when culled, they would be useful for the beef market, suggesting the persistence of demand for old and culled cattle well into the seventeenth century alongside the trend towards a younger carcass.[8]

Markham also gave the only general account of the cattle varieties in England in the seventeenth century.[9] He confined his description to cattle bred in England and said nothing of Scottish, Welsh and Irish cattle, which were already becoming common in parts of England and from which breeding could have occurred. Markham recognised three main types:

(1) Black Longhorns which were found in Yorkshire, Derbyshire, Lancashire, Staffordshire and the dairy regions of Cheshire. He did not mention the Lake counties of Westmorland and Cumberland, but there can be little doubt that the same type was kept and bred there too. Sixty years earlier Harrison had suggested that this breed was the object of a considerable fancy and Markham also commented on the primacy of their aesthetic as well as beefing qualities. He described them as 'of stately shape, bigge, round, and well buckled together in every member, short joynted, and most comely to the eye, so that they are esteemed excellent in the market'.[10]

(2) Pied cattle were found in Lincolnshire and noted for their small and crooked horns. In contrast to the Longhorns, these animals were

described as tall, lean, not apt to put on weight and suitable for labour and draught. Their location in the marshes and wolds nearest to the Low Countries has suggested to many writers that they were a breed of Dutch origin.[11] They were clearly ancestral to the major English breed of the nineteenth and early twentieth centuries, the Shorthorn.

(3) Somerset, Gloucestershire and some parts of Wiltshire were noted for blood-red cattle, similar in general form to those of Lincolnshire. From their ancestral location in the seventeenth century this type probably spread westwards to Devon, north to the Welsh March counties of Hereford and Worcestershire and east along the south coast to Sussex.[12]

No other comprehensive description of English cattle types then appeared until the compilation of the Board of Agriculture County Surveys and the accounts of William Marshall and Arthur Young at the end of the eighteenth century.[13] The dominant variety over most of the country by this date was the black Longhorn, its most famous source region remaining the north-western coast, until Bakewell and his associates in the Midlands created their fashionable sub-breed.[14] The multicoloured Shorthorns, reinforced by further Dutch imports in the late seventeenth and eighteenth centuries, were distributed more widely through the eastern coastal counties than they had been in Markham's day.[15] The red and brown cattle of the West Country were now recognisable as a series of variant types. The original Somerset breed had all but disappeared but the Brown breed of Gloucester-shire was still reasonably common as was the multicoloured Hereford breed slightly to the north. Outlying variants of this western type, completely red or brown, were to be found as the Devon and Sussex breeds, with a cross-bred eastern extension of the latter in the Kentish home-bred, so called to distinguish it from the large number of regional cattle which were brought into that county to fatten in the Weald and marshes. In addition, there were two other breeds to which Markham had made no reference, the small Norfolk, of the same shape and colour as the Hereford, and the polled and ugly Suffolk Dun, the heaviest milking breed of them all. Since its conformation was so bad it could never be developed into a dual-purpose milk and beef breed, so that it was no match for the lower-milking but better-beefing Shorthorn in the nineteenth century. The ultimate fate of the majority of the western, red types in the nineteenth century was to be developed as beef breeds. The only exception was the Gloucester Brown, derived from the original milk cow of Somerset and North Wiltshire, which survived throughout the nineteenth century in small pockets of resistance to incoming breeds – the Longhorns in the late eighteenth century and the Shorthorns in the nineteenth.

Despite the lack of any real specialisation of breed by function in the

seventeenth century and the generally mixed nature of the cattle population, Markham strongly advised against cross-breeding animals of different
types or from different regions. If crossing of stocks was unavoidable, then
the cattle should all be of the same general type, all black races together or all
red races together and so on, and this despite the common practice of cross-
breeding horses, of which Markham himself was a leading advocate. At the
beginning of the eighteenth century John Mortimer repeated Markham's
plea that cattle should not be indiscriminately hybridised, since he believed
that such crossing would lead to the loss of the characteristics for which
each local breed was known.[16] Such crossing was more or less inevitable in
the grazing counties where large numbers of cattle types from various parts
of the country were all run together. The local breeding herds of the region
would be of mixed origin unless considerable efforts were made to keep them
of one type. Even in the heyday of fashionable Longhorn breeding in
Leicestershire in the late eighteenth century, the average grazier there cared
little what type of animal he fed or bred so long as it brought a reasonable
profit.[17] In the mid-eighteenth century, Thomas Hale noted the same
general phenomenon:

the graziers have mixed them [the cattle of different counties] more or less in each
county, though 'tis best for both beauty and service to keep them separate; their
kinds and dispositions being in each particular, as well as the colour.[18]

Farm inventories of the seventeenth and eighteenth centuries show that
colour variants among herds of cattle were the rule. Surviving farm
accounts of the first seventy years of the eighteenth century show the same
mixed-colour patterns, but in some of the larger herds it seems that one
colour type often dominated. At Ombersley, near Worcester, in the 1760s,
the predominant colour of the animals was a red base (43%) with no other
colour amounting to more than 10% of the total. This was perhaps to be
expected in a region so close to the home of the ancestral Hereford, although
the uniformity of red colour in that breed did not reach its present level until
well into the nineteenth century.[19]

There were those in the eighteenth century who advocated the construction of deliberate crosses in order to combine the desirable properties of two
separate varieties in one animal, as had been common practice among
equestrian breeders for some time. Their recommendations were vague and
it is impossible to say whether the authors envisaged a re-creation of the
hybrid from pure stock at each generation, as in the modern, stratified sheep
industry or whether, like the seventeenth-century horse breeders, it was
thought desirable to breed on into subsequent generations from the cross-
bred female stock. William Ellis was an enthusiast for this breed crossing. He

suggested that Scottish and Welsh runt cows, brought down by the drovers to southern England, should be retained and bred from with an English bull to propagate half-breds, which he thought would have the hardiness and milk yield of their dams and the size of their sires.[20]

One of the most obvious changes in the distribution of types between the early seventeenth and the late eighteenth centuries was the greater geographical penetration of the Shorthorn, multicoloured type and the great variety of its local forms. In Markham's day this type was confined to Lincolnshire, but by the end of the eighteenth century it had spread up through the lowland areas of Yorkshire and much of Durham. Yorkshire had been a cattle-raising county in the seventeenth century, but there was no indication that the animals were of any particular note, apart from the Craven Longhorns on the Lancashire–Westmorland border. According to some of the aged informants with whom William Marshall talked in Yorkshire in the 1780s, there had been a brief period of enthusiasm for the Longhorn in the lowland areas in the 1720s and 1730s, but the breed had never gained a strong foothold in the county. The same source told Marshall that before this the general cattle of Yorkshire had been small, black, sometimes white-faced, and of a mixed horned and hornless character.[21] By the late seventeenth century the county already had a reputation for rearing very large oxen, which suggests some of the cattle in the lowlands were already Dutch-origin Shorthorns, which could be grown to a great size.

In the early eighteenth century, North Allerton fair was the major source market for these large Yorkshire beasts, as well as the main staging-post for the Galloways and Highland Kyloes, which had now penetrated to the London market:

they are also good graziers over this whole county, and have a large, noble breed of oxen, as may be seen at North Allerton Fairs, where there are an incredible quantity of them bought eight times every year, and brought southwards as far as the fens of Lincolnshire, and the Isle of Ely, where, being but, as it were, half before, they are fed up to the grossness of fat which we see in the London markets.[22]

The common oxen to the north in Durham were Longhorns, since they were described, by John Laurence in 1726, as having broad, curled foreheads with long horns. But by that date Dutch Shorthorn cattle were becoming fashionable in the county and large prices were being paid for them:

of late years, and especially in the North, the Dutch breed is much sought after and coveted, which have shorthorns and long necks; and they are purchased at very dear rates. Some for fancy (which also governs much the price of horses) will give £20 for a cow of this breed; seeing as they imagine, such beauties and excellencies in them, that they cannot be bought too dear ... the tanner will not give near so much in proportion for their hides; because they are much thinner and lighter.[23]

Marshall's informants believed the main thrust of Shorthorn penetration into North Yorkshire and Durham had been from Lincolnshire and the South Yorkshire region of Holderness, this name sometimes being applied to a local variant of the Shorthorn, which had spread up the Vale of York and reached the Tees valley by the 1740s. The Rev. Henry Berry, writing the history of the Shorthorn breed in the 1820s, also placed the origin of the breed in the Vale of the Tees '80 years ago', that is in about 1750. This breed was certainly based partly on direct Dutch imports and several names of improving breeders have survived, such as Sir William St Quentin of Scampston and Mr Millban of Barmingham. A correspondent to the *Museum Rusticum* in the 1760s said the local Yorkshire cattle had recently changed from a predominantly black to a whitish sort. This almost certainly marked the spread of the paler, pied Holderness or Lincoln Shorthorns among the local black cattle.[24] These Dutch Shorthorns probably never penetrated to the west of the county: Thomas Hale described the normal cattle of Staffordshire and Yorkshire as black, and Staffordshire, South-West Yorkshire and the Pennines remained areas of Longhorn domination.[25] The reputation of the East Yorkshire lowlands for producing large beasts was enhanced by this introduction of the large, pied Shorthorn. When another correspondent to the *Museum Rusticum* published as prodigious the killed-out weight of a bull of the Longhorn type in Staffordshire in the 1760s, a writer from Teesvale replied that the animal was nothing remarkable by the standards of his region, a contention supported by the record of a number of famous giant oxen killed in Darlington later in the century.[26] In the late eighteenth and early nineteenth centuries, when the Shorthorn was firmly established as the local Teeswater breed, considerable numbers of records exist of oxen of huge dimensions.[27]

Apart from its conquest of Yorkshire and Durham, the Shorthorn also had adherents in the South. William Ellis, in Hertfordshire, believed the Holderness breed to be the very finest for the dairyman and veal raiser.[28] Thomas Hale described the Lincolns as large, pied, red and white and very good for beef production, although they would not thrive unless they were well fed. Ellis also commented on their large size and on their 'bags' as desirable milking characters. He mentioned another type of all-white Dutch cow and Hale also distinguished between the well-established Lincoln Shorthorns and newly imported Dutch beasts, regarding the latter type as preferable as milk animals. They were described as long-legged and short-horned, and as being common in Kent and Sussex. An anonymous author of 1767 also mentioned a very high-yielding white breed that could be seen in Lincolnshire and Suffolk and had been brought into Surrey as a curiosity. However, he specifically mentioned Dutch cattle as a separate type of large cattle with high milk yields, given to producing twins. Thomas Pennant, the naturalist, believed that all the larger cattle in the country had come from abroad or were import crosses. He firmly located the Lincoln Shorthorn as an example of this, deriving it from the Holstein. The general picture of the eighteenth-century Shorthorn therefore seems to be one of an old-established red and white import together with paler, better-milking animals of more recent origin from the same source.[29]

Productivity: the size, shape and milk yield of cattle

Literary evidence about the size of early modern cattle is sparse. Harrison considered that English oxen were among the largest in Europe in the late sixteenth century and claimed that the well-fancied Longhorns were 'so tall, as the height of a man of meane and indifferent stature is scarce equall to them'.[30] Other writers merely contented themselves with statements to the effect that English cattle were large. Nor is documentary evidence of carcass weight at all common from this period. The Elizabethan naval purveyors in Devon in the latter half of the sixteenth century specified that the minimum weight of a fat ox should be 6 cwt (672 lb). There can be little doubt that they were referring to carcass, and not live weight, suggesting that fully fatted Devon oxen at this period were probably about 1100 lb live weight.[31] Thorold Rogers provided some analysis of the weights of oxen killed for the table at Winchester College from 1643 to 1688. Again, these figures almost certainly refer to dead weights and represent the carcasses of fully fatted, mature oxen.[32] The mean dead weight was 587.2 lb, corresponding to a live weight of around 1000 lb, slightly smaller cattle than those sought by the navy in Devon.

There was also considerable interest in the late seventeenth century in the fattening of cattle to enormous sizes. Houghton reported that in the 1690s the average price for a large, fat ox at Smithfield was between £10 and £12, but a grazier from Essex had sold a fat ox there for £26 and another from Bury St Edmunds had sold one for £30, which suggests these animals were much above the general size. He gave specific details of another such giant ox, from Newby near Ripon in Yorkshire, whose carcass weighed 106 stone 1.5 lb, which, since these were 14 lb stones, equalled 1485.5 lb. In conformation, this huge beast (two-and-a-half times the size of the carcasses at Winchester) was extremely heavy on the shoulder, suggesting large breeds at this time had been produced primarily for draught. The Newby ox weighed as follows:

Forequarters = 807.5 lb

Hindquarters = 678 lb

The estimated live weight was 2392.5 lb (carcass weight = 62% live weight).[33] Defoe reported that at Steyning, in the Weald of Sussex, Sir John

Fagg had four exceptionally fat bullocks of his own breeding, which he sold at Smithfield in 1698 for £26 a head. One of the purchasing butchers had assured Defoe that the carcasses had weighed 80 stone a quarter, giving a dead weight of 2560 lb, as the Smithfield stone was then 8 lb. This must clearly be an exaggeration since the dead weight would have been almost equal to the live weight of the Newby ox. However, the cattle in Sussex in the seventeenth century were obviously capable of being fatted to a large size, comparable with the largest from Yorkshire and Essex, a reflection not only of the degree of feeding but of inherent size as well.[34]

The Devon and Winchester cattle were small by modern standards. Today, cattle kept under minimal management conditions in the Scottish Highlands mature on the hills at three to four years old and can then be fatted to about 11 cwt or 1232 lb.[35] In the 1930s mature, three-year-old beef was still a major source of supply to the market and the average live weight of these cattle was 1600lb. The specialist 'fancy' ox breeders of the seventeenth century were therefore dealing with mature cattle far larger than those of the average pre-war beef herd.[36]

The other major economic feature to be looked for in cattle was their milk yield. Again, the literary evidence for the seventeenth century is too vague to be of much use. Markham regarded 2 gallons at a milking (4 gallons per day) as 'rare and extraordinary', $1\frac{1}{2}$ as usual and 1 gallon as acceptable. He thought that the sort of cow which could be flushed up to his highest figure would have only a short lactation of three to five months, whereas lesser yielders would give milk for almost the whole twelve months of the year. Worledge thought Dutch cows would give 4 gallons a day and John Smith believed that northern, presumably Longhorn cattle, could also give 4 gallons a day, but only under the very best conditions of pasture growth and grazing management. At Over in Cambridgeshire, on the first flush of April grass, the cows bought in for the dairies of the region would apparently give 3 gallons of milk a day, as would the dairy cattle of the Cheshire cheese region around Nantwich, while on high-season grass in June and July the cows at Over were supposed to yield between 4 and $4\frac{1}{2}$ gallons a day.[37]

With no concrete evidence of how long the lactation of these cows lasted, annual milk yield figures are difficult to derive. Markham's account suggests values of above 300 gallons a year, while Sir William Petty estimated that Irish cattle on their own pastures would yield about 384 gallons annually. Robert Loder's cattle on the Berkshire Downs in the early seventeenth century would seem to have given only about 200 gallons a year, while Sir Cyril Wyche's cows in Norfolk at the end of the century probably did not even achieve an average as high as that.[38] An estimate of considerably less than 300 gallons per head per annum for the national herd in the 1600s would seem to be more realistic than any figure greater than this. It may be compared with the average yield of the national herd in the 1940s, before the modern phase of scientific milk management, of about 500 gallons per head per annum.[39]

Data from the first seventy years of the eighteenth century certainly do not suggest any great change in cattle size and confirm Fussell's opinion, based on an analysis of the literary sources, that the eighteenth century did not see a great increase in the size of cattle as was at one time supposed.[40] In 1717 Jacob Giles gave the following average dead weights for cattle:

Large oxen	900–1000 lb
Middle-sized oxen	about 700 lb
Large bullocks or cows	500–600 lb
Middle-sized cows	400–500 lb[41]

A certain amount of manuscript data about the weights of specific animals between 1720 and 1775 has been analysed[42] and the simple mean weights of different types of cattle carcass were as follows:

Scottish cattle of all types	(13 animals)	315 lb
Welsh bullocks	(3 animals)	569 lb
'Ordinary' cows and heifers	(56 animals)	691 lb
'Ordinary' bullocks	(19 animals)	643 lb
'Ordinary' oxen and bulls	(50 animals)	878 lb
Very large oxen	(7 animals)	1452 lb

If this sparse material can be taken as reasonably representative, it would seem that Scottish cattle were very small indeed but that Welsh 'runts' were not nearly so much smaller than the ordinary cattle of England. Ordinary English cattle appear to have remained much the same size throughout the century and were not significantly larger than those of the seventeenth century. As in the seventeenth century, some oxen were capable of being fatted to sizes well in excess of the average. The late eighteenth-century Smithfield carcass average of 800 lb, quoted by Sinclair, was a reasonable reflection of the general weight of 'ordinary' cattle.[43]

Few dairy accounts have been found for the eighteenth century, so the main evidence for contemporary milk yield must come from the literature. Ellis suggested that a good dairy cow would lactate for nine months and dry off for three months before her next calf. For the first three months, 3 gallons a day could be expected, 1 gallon per day from the second three months and only $\frac{1}{4}$ gallon over the last three months, giving a total annual lactation of about 384 gallons. This estimate was similar to those given in the seventeenth century but considerably higher than seventeenth-century records suggest was actually obtainable. Ellis said the milk yield was sufficient to produce some 200 lb of whole-milk cheese and 100 lb of whey butter, figures that are just about compatible with this yield, assuming the milk had a high butter fat content.[44] If Ellis was referring to Longhorns this would be reasonable, as the linear descendants of this breed still produce milk with a high fat content compared with that given by modern Shorthorns or Friesians.[45] Hale also looked for the highest yield from a cow in the first months of lactation on the spring flush of grass. The 'natural' yield of the cow should be judged at this time and no cow giving less than $2\frac{1}{2}$ gallons a day should be kept. Like Ellis, he regarded 3 gallons as a reasonable norm, although he said rather optimistically that such a cow would give this yield for the whole year, producing an annual lactation of over 900 gallons, hardly a realistic figure.[46] Elsewhere he suggested an expected annual figure of 400 gallons. When discussing butter yield he gave an estimate of 1 lb of butter from 1 quart of cream, obtained from 6 quarts of milk, values similar to those given by Marshall and Young later in the century. Elsewhere he suggested 400 gallons of milk would give 2 cwt of butter, as did Ellis as an alternative to producing whole-milk cheese. (His estimate was 200 lb of butter from 384 gallons.) Both these latter figures seem exaggerated.

The 'Country Gentleman' thought that the smallish black cattle would yield about 2 gallons a day for ten or eleven months of the year, giving another overestimate of 600 gallons or more. He believed Shorthorns would give up to 6 gallons a day but considered that they only had very short lactations.[47] Many authors were guilty of extrapolating the best early spring yield of the cow over the whole year, only Ellis conceding that the maximal yield would not last long. Indications from the butter yield patterns of herds of cows in the late eighteenth and early nineteenth centuries show that milk production did indeed fall as Ellis suggested.[48] In a small herd of Staffordshire cattle in the late eighteenth century each cow appears to have produced $2\frac{1}{2}$ cwt of cheese per year on average. If the 1 lb of cheese to 1 gallon of milk ratio is applied, this gives an annual yield of just under 300 gallons. This seems far more realistic than the literary estimates, although rather greater than seventeenth-century yields so far assessed.[49]

Cattle breeding in the seventeenth century: the unique position of the Lancashire Longhorn

Most literary advice on the management of cattle and the selection of breedstock in the seventeenth century was drawn from classical sources. Many of the features to be looked for were 'fancy' rather than useful points and much of the conformational advice was so vague as to be virtually useless. Bulls were supposed to be fed up before service while cows were to be kept short of food, both before bulling and to some extent during pregnancy and lactation. The flushing-up of the bull would enable him to perform his function as quickly as possible, but the tendency to starve the cow was clearly counterproductive to good calves or a good milk yield. The justification generally given was that the overfat cow would suffer from fat blockage of the milk pores and thus yield would be reduced.

Breeders who agreed with Gervaise Markham would have paid considerable attention to the choice of their bulls for the practical reason that the bull influenced the quality of far more offspring than the cow and the philosophical principle that the male was more significant in reproduction than the female. Markham expressed these ideas like this:

For as much as the male of all creatures are the principall in the breed and generation of things, and that the fruit which issueth from their seed, participateth most with their outward shapes, and inward qualities, I think fittest in this place, where I intend to treat of horned cattle and neat, to speak first of the choice of a faire Bull, being the breeders principallist instrument of profit.[50]

It does not seem that this advice was generally followed since it did not lead to any increase in the monetary value of bulls in relation to the product and profit-yielding cows. An analysis of a series of published inventory collections was used to assess the relative values of bulls and cows.[51] These collections show a bias towards the northern counties, where cattle breeding was becoming more significant than it had been, as stratification in the cattle trade grew. If bulls were valued for their breeding function, this should show up in northern inventories. However, from a total of 117 usable entries, only 12 had a ratio where the bull was more valuable than the cows, and that included several where the difference was marginal. In the great majority of cases the bull was less, in many cases substantially less, valuable than the cows, while in about half the entries, the bull and cow valuations were either the same or the types were not distinguished.

In the mid-seventeenth century Samuel Hartlib, one of the enthusiasts for the husbandry of the Netherlands, commented unfavourably on the laziness of English breeders in failing to select out strains of cattle for milk or beef specialisation and persisting with general-purpose animals. Ultimately such specialisation was to be one of the keys to cattle improvement in the nineteenth century. Such specialist breeds imply specialised niches in the agrarian economy and it is doubtful whether these existed on a wide enough scale before the late eighteenth century to justify this approach to farm livestock improvement. Hartlib spared one area from his castigations:

We are too negligent of our kine, that we advance not the best species; for some sortes give abundance of milke, and better than others: some sorts are larger, more hardy, and will sooner fat ... Lancashire and some few Northern Counties are the only places, where they are a little careful in these particulars.[52]

Lancashire was the home of the Longhorn, already subject to a considerable fancy in the sixteenth century. Some account of the way in which seventeenth-century breeders were a 'little careful in these particulars' may be gleaned from correspondence in early agricultural journals published by John Houghton in the 1680s and 1690s.

The first Lancashire correspondent in 1683 was a Mrs E.H.,[53] a dairywoman of Hallsall, a parish near Ormskirk on the pasture zone of the Western Lancashire Plain. The principal object of breeding was the maintenance of a dairy of cows, with the heifers bred up as replacements and the steers used as draught oxen on the local farms. Most of her account dealt with such a dairy herd and the rearing of calves. The optimal age for a milk cow was reckoned at six years, and she should be bulled by a broad-headed, large bull of about the same age. Although this was a Longhorn region, the cows were expected to have small or slender horns, with a tail which reached down to the back knee. Mrs E.H. did not mention any other obvious criterion for selection, such as milk output or udder characteristics and Houghton commented on this with some surprise in notes appended to this letter:

The description of a broad-headed, slender horn and long tayl ... I set down as I receive it, that the curious observer may confirm it as the truth or detect the error; but the country folk value their cattle by these marks much.[54]

The cows calved in late March or early April and the calves suckled for a month from their dams. Then they were removed from the milking cows and put to old nurse cows of 15 or 16 years of age, whose active milking life was over. The calves were then weaned off these nurses in August and wintered on hay and oat straw. They were put to a mature bull in July or August when they were more than two years old, although a younger three-year-old bull was used on these heifers, rather than a large six-year-old. At about Christmas time the young cow was tied up indoors for shelter and taming before the spring calving. The three-year-olds were now part of the dairy herd and would go on milking until they were twelve to fifteen years old, when they were pensioned off as nurses for three or four years.[55] After that the old cows were bulled and fatted for beef.

In the early editions of his later *Journal* Houghton ran a series of editorials on cattle and included a reprint of Mrs E.H.'s letter. This discussion prompted another Lancashire correspondent to comment on the description of Lancashire practice, from Hallsall.[56] He agreed that milk cows were in their prime at between five and eight years old but believed there were no more than two six-year-old bulls in the whole of Lancashire. Those who were curious in their breed made use of one or two-year-old bulls, while bulls of four years old and above were so slow and heavy that they could do little or no service. The young bulls would serve a pasture of some forty to sixty cows and heifers in the summer season. Nowhere was this stated, but the commercial premium of the young bull probably lay in his ability to service this number at pasture in the shortest possible time, so that consequent calf drop was closely bunched and all cows were well into their lactations once the spring and early summer grass growth accelerated their milk production to maximal levels. This premium was reflected in price differentials, since £10 to £12 would be given for a promising yearling bull at Inglewood fair, while the same bull when four years old might be bought for £4 or £4 10s as 'not being fit for the service of breeding' and would then be gelt and worked for a couple of years before being fattened. This was the first clear literary statement of an enhanced value being given to a bull solely as a consequence of his breeding capacity, a reversal of the traditional absence of any real commercial value for such a beast. To achieve the onset of sexual maturity in a yearling involved pushing his growth and development extremely hard, an expensive process which was justified by the high value of a good yearling bull.

The best such potential bulls were born in March, the progeny of parents with the best general phenotype, vaguely described as 'fine', 'large' and 'young'. They were treated like all the other calves for the first month and allowed to suckle freely from their dams. For the following three months each selected calf was given a nurse cow to itself. This first was then exchanged for a second nurse for another two months, and a further nurse substitution made for the next two months. By mid-November this cow and the calf were night-housed and the calf had hay and oats as well as milk. By Christmas another fresh nurse cow was produced and the calf ran with this cow until April when the bull calf would be 13 months old and larger than the nurse. By July the young bull was deemed fit for service and ran with the cows in the pasture. Such beasts were said to be generally easy to handle, except in the mating season itself when they became almost uncontrollable and ran bellowing from field to field. If two bulls met they would often fight.

The picture painted here resembles Greek practice in Epirus during the classical era. The cows were run in the pasture for service and the bull allowed to rampage freely among them, a procedure which cannot have been very safe, although the extreme youth of the bulls may have vitiated some of their aggression. Details of how these large dairies were milked were not given and how milking was achieved in the month or so that the bull was among them (June and July for the April calving on the late-growing northern pastures) was nowhere specified. The high price of such yearling bulls, either purchased or reared on the same dairy (although the latter was unlikely given the prejudice against 'continuing in the same breed'), can only have made the exercise worthwhile for the large dairy owner, who must have constituted a small minority. Putting a calf to several nurse cows was an extravagant procedure, only likely to be used on the larger estates. The latter point was confirmed by a third writer from the Lancashire area, who described the simpler practice of the smaller cowkeepers.[57] This last author thought that only 20% of Lancashire calves were reared on nurse cows, while the great majority not only had no nurses, but were removed from the teats of their dams after seven to ten days and not allowed to suckle for a whole month. They were taught to drink milk from a pail and at the end of the first month, the whole milk was increasingly diluted with 'pottage', a mixture made of blue milk (the first drawings of each milking), skim milk and oatmeal. This was standard practice for small cowkeepers and cheese specialists, with a large supply of skimmed whey, from May onwards. Four reasons were given for the adoption of this latter system: that dams over-familiar with their calves would pine when they were withdrawn and be reluctant to give milk, that whole milk could not be spared from production for profit, that small cowkeepers did not have any spare cows for nurses and that the earlier the calf was removed from the cow the more readily she would accept the bull in May or June, so that she would be well in milk to take advantage of the earliest grass growth in the following spring.[58]

These descriptions suggest a stratified local industry where the larger dairies operated as major breeding concerns as well as milk-producing units and had the resources to use complex techniques which they believed would lead to high-grade cows and early-maturing bulls. This type of animal was instrumental in impregnating as many cows as possible in the shortest possible time. Given the practice of running the bull out with the cows at pasture, they had no option but to go for a bull that would impregnate the whole herd, rather than use two or three, which would invariably fight under these primitive conditions. Only a very young bull could do this in a short time. To those practical reasons for encouraging juvenile bulls could be added the feeling that the physiological vigour of the male would be transferred to his offspring. The sluggishness of old bulls and boars had been noticed and allowed for in ancient Greece and the same views seem to have been evident here but carried to a logical extreme, with very young stock being used, presumably on the grounds that they would have more activity and 'lust' than even prime-age bulls of three to five years. Elsewhere Houghton noted that bulls should be employed before they had acquired full size, but at the age of two or three years, rather than as yearlings.[59] Husbandmen around London bought bulls as two-year-olds, used them for a season or two and then sold them off fat.[60] But in contrast to the premium prices paid in Lancashire, they bought their bulls at £2 10s or £3 and sold them fatted at £5 or £6, the traditional differential between store and fat animal prices.[61]

There was therefore nothing exceptional about using a non-mature bull, but the prices given and the lengths resorted to in Lancashire to push the sexual age down do appear to have been unusual. One of Edward Lisle's neighbours in Hampshire suggested the use of yearling bulls on the smaller upland cows of the region for the practical reason that a mature bull would damage them.[62] Lisle also reported a tendency towards breeding from very youthful cows as well, especially in Somerset, where two-year-old heifers bearing calves were apparently not uncommon, although it was believed by many that this stunted the dam's growth.[63] Another of Lisle's neighbours commented that it was better to rear heifer's calves to provide the next generation of milking animals than calves of cows. He provided the practical explanation for this preference by pointing out that the heifers were not milked at the same time the calf was going through early foetal development; consequently, it was not 'robbed' of nutrition and would therefore develop more fully in the uterus.[64]

No clear picture of breeding objectives or techniques emerges here. Most attention was given to them in the dairy regions but this was largely governed by the practical considerations of lactation timing. As techniques of selective breeding Lancastrian methods were defective since no consideration was given to economic characters in the phenotype. The bulls chosen must have been selected at a very early age since no more calves would have been run through the expensive nurse system than were absolutely necessary, and there was no hint of any formal progeny-testing procedure or examination of pedigree. However, high-plane nutrition among the larger herds for both bulls and cows must have produced a physiologically improved strain and the policy of juvenile breeding may have led to a number of accidental selective consequences, such as the build-up of genes for fast growth rate and early maturity in the top stratum of the Longhorn breed. Any such accidental shift in average genotype would require these policies to have been pursued consistently by a group of breeders for a long time. If this was the case in the Lancashire and Cheshire dairy regions it might help to explain the superiority of the north-western Longhorn and its adoption by Midland Plain breeders, including Robert Bakewell, as an animal for attention in the mid-eighteenth century. Certainly at the end of the seventeenth century, the superiority of Lancashire cattle over anything raised in Leicestershire was already obvious.

Edward Lisle had a number of relatives and contacts in that county in the early eighteenth century and he made observations on the husbandry of this region as well as his native Hampshire. His two chief sources were a Captain Tate of Loughborough and a Mr Clerk of Ditchley, and in 1706 he conversed with both of them on the subject of the widely imported Lancashire Longhorn. Tate expressed surprise that, despite the richer Leicestershire pastures, imported Lancashire cows and calves would degenerate, so that by the third descent they were more like the common Leicestershire stock than fresh imports. Clerk supposed the main reason for the difference to lie in the relative care taken over breeding in the two counties. He reported the

Lancashire practice of using yearling bulls, the premium prices they commanded and the use they made of unskimmed milk to feed their calves, where Leicestershire breeders would use skimmed milk and whey after only a month on whole milk (the same procedure employed by the smaller Lancashire men). He believed Leicestershire dairymen, who kept only small dairies, would not go to the same trouble because their pastures were more profitable if used for fatting cattle and breeding coach horses. Dairying was only a minority pursuit.[65]

Care over cattle breeding may have been confined to dairy regions, where the milk cows were retained and the same animals used both for breeding and for the economic yield of products. In the purely breeding zones such as Ireland, south-western Scotland, Wales and the Scottish Highlands, any conscious breeding strategy directed towards the end-product, the beef animal, probably did not occur because the unfinished animal was disposed of to the drovers for transport to England. From the point of view of the grazier this may have been no bad thing. In terms of beef conformation, essentially feral animals, such as those bred in south-western Scotland, or the Kyloes of the Highlands, seem to have been base stock superior to any of the breeds which had been extensively bred for draught or milk purposes for centuries in England. Thus Bakewell was reported by Sir John Sinclair in 1825 as having said that he wished he had used Kyloe cattle as a basis for a beef animal since they were perfect in all respects except size.[66] William Marshall also remarked at the end of the eighteenth century that the feral White Park cattle of Northumberland were perfect in beef conformation.[67]

Lancashire had long been a region specialising in cattle breeding, but a change from a rearing to a dairying economy can be detected at the end of the sixteenth century. The same transition to milk production from simple breeding occurred in other, not too peripheral, breeding regions of England at the same time.[68] The system of using high-priced yearling bulls in Lancashire may have originated in this transition to dairy farming. The detailed farming accounts of the Shuttleworth family at Gawthorpe Hall near Burnley provide some hint of this connection.

The steward's accounts for the family run from 1582 to 1621.[69] The purpose for which their small herd of cows was kept is unknown, but it did not seem to generate sufficient replacement oxen since these were often bought in. There were no purchases of cows recorded after 1592, so that one could envisage a small, closed breeding herd based on the cows and heifers bought between 1583 and 1592. Sporadic sales of cows and heifers took place until 1620, when 18 heifers were disposed of. If these accounts contain all the sale and purchase transactions, it suggests that the herd size grew from half a dozen or so in the 1580s to perhaps more than twenty by 1620, mainly through in-herd replacements. Bull purchases were irregular and showed no consistent pattern or policy. The fact that the one bought in 1594 was specified as a two-year-old suggests that unspecified bulls were over three years old or mature animals. The beginning of what seemed to be a policy of regular biennial purchases can be detected between 1602 and 1606 but this was not kept up.

Where bulls were bought and cows and heifers bought and sold in the same year, bull – cow value ratios can be estimated. The results are highly variable and show no consistent over- or undervaluation of bulls relative to cows. If the values of all the cow transactions and bull purchases are averaged out, an overall ratio of 1.254 is obtained, showing that bulls were generally more valuable than cows. Price differentials, such as those of the late seventeenth century for yearling bulls in the same region, did not occur. But there is a hint of a trend in this direction. Prices for cows and heifers show no wild fluctuations about the mean price of £2.47, but the last two bull purchases in 1611 and 1617, with prices of £4 15s and £6 respectively, were well outside the norm ranges of any cow or bull value obtained before. These are only two values, but they point to the possible rise of specialist bull-rearing interests in the Lancashire Forest cattle regions. Cause for doubt about the reliability of these data must also be raised by the failure of the general price for cattle here to conform to the expected price inflation during this period.[70]

Cattle breeding in the eighteenth century: the spread of the Lancashire breed and Lancashire methods

At the beginning of the century Defoe described the stratification of the trade, which was much as it had been in the seventeenth century.[71] The eighteenth century saw more discussion of the veal and dairy regions to the north of London in Essex and Hertfordshire, where considerably more trouble was taken to obtain suitable breedstock than had been the case previously. Lancashire retained its fame as a major source of well-bred stock throughout the first half of the eighteenth century, but thereafter the graziers of the Midland counties began to master some of the techniques used in the North-West and an elite group of Midland breeders appeared. These men, in turn, were displaced as leading exponents of the art during the early years of the nineteenth century by Shorthorn enthusiasts in Yorkshire and Durham, and Hereford breeders in the South-West.

The literary and manuscript evidence between 1700 and 1770 can be used to show that two things seem to have happened. The first was that the techniques employed by the Lancashire breeders diffused to other parts of the country, and not only to the Midland counties,[72] and secondly that agricultural writers and animal breeders were becoming more and more convinced that the desirable properties of any type or breed were as much hereditary as environmentally determined. They began to believe that the application of selective breeding methods could lead to improvement and that the transfer of breeds from one part of the country to another would not necessarily lead to their degenerating or changing into those of the locality to which they had been moved.

John Laurence, who had recorded the rising fashion of keeping Dutch cattle in the 1720s, also noted the Lancastrian habit of keeping cows to act as nurses to high-bred stock and suggested that the same procedure had spread to the Essex veal producers.[73] John Mortimer also recorded that in the breeding areas, which he did not specify, calves were allowed to run with their dams rather than being removed.[74]

Like the dairyman, the producer of veal calves was particularly concerned with the milking properties of his cows since the calf lived mostly off its dam's milk over the major part of its short life.[75] Both Houghton and Laurence gave some account of their procedures,[76] but it was William Ellis who gave the earliest comprehensive description of Essex and Hertfordshire practice.[77] Although he had recommended the Holderness Shorthorn as a suitable cow for veal production because of its reputation as a milk animal, he was aware of the fact that large size was not necessarily related to yield and, indeed, that on the Chilterns, where forage was never plentiful, a small animal would probably give more than a half-starved large one. More clearly than any of his seventeenth-century predecessors, Ellis also saw that one phenotypic key to the choice of a good milk cow would be the size of its 'bag' as he called it. He always looked for a 'good bag' in a milk cow, and felt for the presence of large milk veins going to the udder. Apart from milk yield, the veal raisers wanted an animal that would have the requisite pale flesh, to which end they recommended various chalk-based diets and a range of more devious tricks to ensure the whiteness that the consumer wanted. Ellis believed that some external marks were indicative of white fleshing, so that the best veal producers looked for bulls with pale eyes, pale skins and pale inner ears as getters of calves with pallid flesh. They must have been one of the few groups of breeders in the country who preferred pale animals, for most were prejudiced against them because of their lack of hardiness.[78]

This belief in the relationship between the colour of the bull and the colour of the flesh of the calf led to considerable effort to find breeding bulls with the correct characters. Some were 'so curious' that they would ride fifty miles in search of a suitable specimen, going especially to the fairs in Huntingdonshire. Apart from the pallid animals which were their favourites, they would also use black bulls with brown finch marks on their backs and deep-red animals as well. The bulls were usually two-year-olds or yearlings, three- and four-year-old animals seldom being used. Ellis reckoned that such a

Plate 11. The Longhorn bull, *Shakespeare,* bred by one of Bakewell's contemporaries, Robert Fowler of Little Rollright, Oxfordshire. This picture by an unknown artist is thought to have been painted in about 1790. *Shakespeare* was one of the most famous improved Longhorn bulls. The animal passed from Fowler to Thomas Paget in 1791, when Fowler's stock of 50 cattle were auctioned. He was resold at Paget's own stock dispersal sale in 1793 for the record price of £400. Photograph from the Museum of English Rural Life, University of Reading.

yearling could easily serve a group of twenty cows, rather fewer than Lancashire breeders had expected fifty years earlier. Bulls became difficult to handle when they were mature and Ellis believed that using polled animals would save a great number of accidents. Ellis had noticed the mothering qualities of Scottish runts and was beginning to favour these as dams and nurses to the calves. Laurence's reference to nurse cows in parts of Essex probably derived from the veal trade and not from the use of nurses to rear breedstock, for though the Essex veal men used yearling bulls, they did not breed these themselves, nor did they perpetuate their own variety. But on the rich pastures of the South Midlands very young bulls were evidently available by the 1740s and 1750s.

Reflecting the emphasis placed on the dairy by advanced breeders in the early eighteenth century, the extensive breeding advice given by Thomas

Plate 12. A long-horned Lancaster cow bred by Fowler of Little Rollright. A mezzotint after a painting by James Ward (1769–1859). The mezzotint is also probably by Ward. It was not published until 1820, almost 30 years after the dispersal of the Rollright herd. The painting may date from after 1800 and be one of the unfinished series of 200 paintings of British breeds of livestock commissioned from Ward. Photograph from the Musuem of English Rural Life, University of Reading.

Hale concentrated on milk properties. Hale believed the bull was more important in imparting conformation to the ox than the cow, clearly relating the sex of the parent to the control of the form of offspring of the same sex. His advice on the desirable shape for a bull relied heavily on classical precepts of beauty and included the broad forehead, curled hair, moderately long horns, thick and fleshy neck, big breast, large body, flat, straight back, large, square buttocks, short, straight legs and so on. In addition, he noticed, as had Houghton, that many country breeders also placed great store by certain marks 'of less consequence, tho' some stand upon them with great strictness'. There was no obvious rational basis to the selected marks, as there was in the case of the Italian horse colour rules. They included roughness of the inside of the ears, wide nostrils, a long, thin and hairy dewlap, a high-standing tail, large, round knees and several others.

Somewhat confused in his mind about the relative contributions of male and female, Hale finally chose to emphasise the male in breeding: 'the owner can never be too particular in his examination of the bull which is to be the father of his breed'.[79] This advice to look to the bull echoed Markham, but it still did not lead to many, apart from those in Lancashire and later Leicestershire, being sufficiently careful in their choice to produce any general elevation of the value of bulls. Farm accounts in general for this period show no excess valuation of bulls over cows, any more than the inventories of the seventeenth century did,[80] so that even Hale believed the bull should do something else to justify his keep and offset his costs. He therefore advocated that bulls should be worked like oxen and claimed that in many places this was done successfully.[81]

Whatever breed was chosen, the cows should be notable for their milk: 'Above all things see that she have a large, good, white and clean looking udder with four well grown teats.' Since Hale was an advocate of pure breeds he recommended that whatever breed was selected for the cows, the bull should be as near as possible of the same type. He also expressed the firm belief that cows of any county could be raised in any other, implying that he thought the main breed characters fixed and hereditary, and that the breed would not degenerate or change when transferred. It may have been the growth of this conviction that persuaded some Midland breeders to work on their own Longhorns rather than constantly import them from the North-West.

The breeder should, according to Hale,

take care to have a right bull, and well sized cows, both of the same breed; and from this stock, with a sufficiency of pasture rich enough for their support, he will not fail to have either breed in any county: perhaps better than they are to be found in the places themselves, as many do not take due care there, but bring in one kind among another.[82]

Although the examination of appearance was all that could be judged when an animal was purchased, this in itself was not adequate, and every cow in the herd should have its milk yield noted. The recording system was based upon the yield that the cow gave immediately after calving, on the grounds that the yield would then be maximal. Hale did not apparently notice that the length of lactation and the yield throughout that time would need to be considered in a proper analysis of milk yield. Nevertheless, he had stated the basic principle of building a herd by performance testing, although he had not devised an adequate measure of that performance. Every cow in the herd was to be monitored and any animal that did not come up to a certain standard should be culled out and sold without delay, on the chance that the new purchase to replace it would do better. The minimum standard that he believed acceptable was 10 quarts ($2\frac{1}{2}$ gallons) per day.

Anything with a lower performance was culled. A cow was therefore maintained in the herd provided its initial yield was better than acceptable and only the performance of the marginal cows in the herd was monitored carefully. Hale's hereditary view did not extend so far as recommending that herd replacements could most usefully be bred from high-performance cows.

There was no suggestion here, as in the case of the seventeenth-century equestrian authors, that once an initial group of females had been decided upon or assembled, they should constitute a closed herd. He confined himself to the view that sale, exchange and purchase were the ways to build up the herd's overall performance:

This is an essential point, for if the difference in a single cow be something, as it certainly is, 'tis very great in the produce of the number, and this is a certainty that the quantity of milk does not depend upon the quantity of food, though the quality of it does often on the nature of the food. A cow that is starved will not yield much, that stands to reason; but if a couple of cows be watched, one of which is deep in milk and the other not, she that is the least profitable, will be found to consume as much provender, let it be what kind it will, as the other.[83]

The same general point about assessment for milk performance and the fact that it was innate was made by the 'Country Gentleman'. He had observed that marsh-fed Lincoln cattle gave far richer butter than the same animals fed on short grass, but nevertheless he was aware that some breeds would never give a high butter yield, even under the best conditions of nutrition, while others would give a good yield of butter even on the poorest pasture:

Again the milk of one cow shall give richer and better butter than the milk of others, though they all feed on the same pasture, even so that the milk of one cow will cover or enrich the butter made from nine or ten other cows; her milk may make butter of a rich yellow colour, full of fatness, and the others will only produce a pale, lean butter, but all together will be good.[84]

Only one agricultural writer before 1770 seems to have extended this principle of heredity to its logical conclusion of selective breeding to improve the standard of the whole breed up to and beyond that of the best currently kept. This was an anonymous correspondent from Teesvale in 1766, writing about the large size and high yields of the sheep and cattle in the vale. He clearly stated the benefits to be obtained from the careful selection of breedstock for any character thought desirable, which in his case was still large size as the best indication of productivity. He was not completely sure that inheritance was wholly isolated from the environment and felt that its reinforcing features were important. Nevertheless, he believed the properties of animals from different geographical locations would be retained when they moved to new areas, much as Hale had argued, but emphasised in addition the need for selection among the best individuals for breeding:

Plate 13. A Staffordshire–Longhorn cross heifer owned by Bakewell's contemporary, Thomas Princep, and painted by James Nost Sartorius (1759–1824). When and why this particular animal was painted is unknown, but presumably it dates from the 1790s or early nineteenth century. Photograph from the Museum of English Rural Life, University of Reading.

In almost all instances of peculiar properties and qualities in the individual, the progeny, for the most part, more or less, partake of the same, though in very various degrees, and some particulars not at all. If, therefore, those of them, in which such qualities are most predominant, be successfully selected for breeding, the qualities themselves will be greatly augmented, even to a very high degree, in many cases. This principle extends as well to form and magnitude, as to dispositions, passions and other qualities . . . The peculiar soil and climate of many places, adds much to this in many cases: and the qualities imputed to the breed, by the effect of soil and climate, will frequently continue in the race for a considerable time.[85]

The statement that selection would apply to improvements in productive form and size as well as features of personality or character suggests that the model for this process of hereditary transmission was anthropomorphic, or derived from horse breeding. As in antiquity, it seems that the stimulus for animal improvement came from the conventions (many of which were confused with social and political issues) of hereditary transfer in human society.

Contemporary with the appearance in print of this faith in heredity and the value of selective breeding, Midland graziers in Leicestershire were developing imported Lancastrian Longhorn cattle into something of a fashion. It seems clear that Bakewell and his Midland colleagues selected the best Lancastrian stock they could find and adopted Lancastrian breeding methods, which had been diffusing slowly down through the Midlands during the early eighteenth century. For instance, in 1751 the tenant at Coton Hall Farm in Shropshire paid 1s each for his own cow and that of one of his neighbours to be served by Lord Stamford's bull. The size and hide weight of one of Stamford's bulls which was fatted at Coton in 1763 suggests that these bulls were Longhorns and that Stamford may have been importing them from Lancashire. This supposition is strengthened by the fact that a cow specifically designated the 'bull's nurse' was also killed and another nurse sold. The carcass of this animal weighed only 360 lb compared with the 909 lb of the bull, and it was probably a small Scottish beast.[86] In the Midlands in the later years of the century, the nurse cows for bulls were often not Longhorns, but Scottish or Welsh runts.

The Drake family, in Buckinghamshire, had considerable out-county farming interests at their estate at Ombersley, five miles north of Worcester. Their dairy herd in the 1750s and 1760s was composed of nine or ten animals and thought large enough for the upkeep of a bull to be justified. In fact, in most Michaelmas stock lists two were recorded, the service bull and a yearling bull which would replace it, since the two-year-old only covered for a season before being fatted for disposal. There is a record of a nurse cow in the 1761 stock list in addition to the nine cows in the dairy. This animal could have been maintained as a nurse for a young bull, in the Lancashire fashion. Among the 1762 calf drop, two of the six calves were sired by a white bull which was still only a yearling when he had served, suggesting that both major elements of the Lancashire system had spread as far as Worcestershire by the 1760s.[87]

The Dixon family of Revesby and Firsby in Lincolnshire kept a small dairy of between five and ten cows in the latter half of the eighteenth century. The bulls used to service this herd were usually listed and were nearly always animals from outside owners. From 1759 to 1763 the bull used for all cows was described as Ostler's bull, although it was not always the same animal. But in 1764 a Lancashire bull was specified for repeat services in the second six months of the year and in 1765 a Lancashire bull was used until August, when matings reverted to the old bull. In 1769 they again used what was called a 'New Lancashire bull' together with a young bull, but the Lancashire was not serving in 1770. In both cases the Lancashire animal was only used over a twelve-month period, whereas local bulls were often used for two seasons. Some of the cows in their herd, which was firmly located in Lincoln Shorthorn country, were also described as Lancashires, never more than one until 1764, but between 1765 and 1770 there were between three and four in the herd, which varied in size between eight and ten cows. But from 1771 the number declined to one again. It looks very much as if the Dixons decided to try out the Lancashire beasts, which were acquiring a reputation in Leicestershire and Staffordshire, but had decided, after experimenting, to stick with the Shorthorn breed of the locality.[88]

Among the many notable breeding graziers of the Midland counties recorded by Arthur Young on his eastern tour was Sir Robert Burdett, of Formark in Derbyshire.[89] He had a particular herd of polled cows, chosen because they would not damage young plantation trees, but they were not such good milkers as the Longhorns of the region, one of which Sir Robert had got to produce $4\frac{1}{2}$ gallons a day of good quality milk at peak lactation. His accounts show that his stock on the farm in January 1765 consisted of 14 Welsh bullocks grazing with a dairy herd of 7 of his polled cows, 1 black cow and a heifer purchased in July 1764 specifically as a nurse cow for his young bull calf, presumably of the polled breed.[90] Once again there is evidence that the Lancastrian habit of running a selected bull calf on the teat had spread down into the Midlands.

This selection of farm accounts suggest that advanced tenets of cattle breeding were reasonably common in the Midlands in the 1750s and 1760s. Legend has it that Robert Bakewell was the greatest of the Midland improvers and it is true that when Arthur Young made his first visit to Dishley Grange in 1770 Bakewell, who was then about 45 years old, already had a national reputation.[91] William Marshall reported that Bakewell and his most famous Longhorn-breeding contemporaries, Princep, Paget and Fowler, had all drawn heavily from the herd of an earlier improver, Webster, of Canley near Coventry, who had been the leading breeder on the Midland Plain in the 1740s and 1750s.[92] The names of two of Webster's contemporaries as noted breeders have also survived, Sir William Gordon and Mr Phillips, both from Garrington near Loughborough.[93] Further back along the pathway of Midland foundation stock lay Sir Thomas Gresley, of Drakelow on the Trent in Staffordshire, from whom Webster apparently drew his foundation stock of six cows in about 1728, as did Welby of Linton, another shadowy figure, whose herd was supposed to have achieved considerable fame before it was destroyed by distemper.[94] As with the annals of the Thoroughbred, certain foundation animals and their owners or breeders could be traced in the ancestry of the leading exponents of the Midland Longhorn breed, once it had become well known. These breeders had their names recorded almost by accident, and subsequently it seemed

Plate 14. The *Blackwell Ox.* A Shorthorn ox killed at Darlington in December 1779. It was bred and fed by Christopher Hill at Blackwell, County Durham, and its picture was painted after its death in 1780. The print here reproduced was not published until June 1809. The animal was bred for beef and its measurements show that it was of improved form with equal forequarter and hindquarter weights. This animal was representative of the type adopted by Thomas Booth and the Colling brothers in their successful improvement of the Shorthorn breed which eclipsed the Midland Longhorn in the early nineteenth century. Photograph from the Museum of English Rural Life, University of Reading.

that they operated in splendid isolation as enlightened breeders, head and shoulders above their contemporaries. But the Midland farm account data suggest there were a number of breeders and graziers across the area, all importing north-western stock and ideas, some of whom were enthusiasts and fanciers, who were prepared to spend time and effort on cattle in the same way that the gentry were prepared to lavish attention on racehorses. Despite his enthusiasm for Bakewell's stock and husbandry, even Young did not fail to notice the extensive fancy for Longhorns throughout the Midlands and Lancashire. When Young visited Dishley in 1770, Bakewell's lead bull, *Twopenny*, was covering cows at 5 guineas a service and he had others let for the season at between 5 and 30 guineas. These were certainly exceptional prices but they probably say more about Bakewell's undoubted gifts of salesmanship and self-publicity than about his stock. More germane for comparative purposes were the sale valuations and prices of his stock.

Young noted that Bakewell would not sell some of his best breeding cows, even for 30 guineas each. A Mr Knowles of Nelson near Loughborough had 'dairy cows which he values at £20 apiece, and some that he would not take £30 for'. Mr Mundy, near Derby, had sold cows for £25 a head, while Charles Turner, of Kirkleatham near Gisborough, had found great difficulty in acquiring base stock upon which to start his Longhorn herd and had paid 20 guineas each for 15 cows. In Lancashire around Garstang the best cows of impeccable Longhorn parentage would also sell for up to £20 or £30 a cow. In the same region, good bulls could apparently command up to £200. Knowles had imported a Lancashire bull from its home region and had paid 60 guineas for it, while Charles Turner had had to pay 40 guineas for his first bull. These prices should be seen in relation to those for ordinary Longhorn grazier's stores, which were apparently sold for £6 to £10 around Hazelbeach in Northamptonshire for Shropshire-bred beasts, and about £7 to £10 around Kettering for cows for the dairy or fatting.[95]

The pattern of development in the Midlands seems reasonably clear. A superior version of the Longhorn breed, originally confined to Lancashire, Westmorland and Craven, and from the late seventeenth century constantly reintroduced into the Midlands, had been transferred to many of the Midland counties as breedstock by the mid-eighteenth century. Locally acquired or perhaps imported herds of cows were crossed against bulls constantly obtained from the north-western source regions or, in the later years of the century, from the best herds of the neighbouring, locally raised cattle. By then, further recourse to the original source was not necessary because the local Longhorn sub-breed had been transformed, by direct import and grading-up, into cattle as good as those in the original location. An almost exact parallel with the development of the Thoroughbred may be seen, except that the foreign source for grading-up was very much more distant in the latter case. Before 1770 a return to the base regions for male breedstock still appeared to be common. Bakewell's *Twopenny* was bred from a Canley cow (known as *Old Comely*) which had calved in 1765, crossed with a bull imported from Westmorland.[96] Some of the most famous animals bred by Webster himself had also been produced from the same combination. One of his bulls, *Bloxedge*, bred in the 1750s, whose most notable daughter was the foundation dam of Princep's Croxall herd, had been bred from one of his own cows crossed with a bull imported from Lancashire by one of his neighbours.[97] Confirmation of the universal spread into the Midlands of the essential features of Lancashire technique by the late eighteenth century came from Marshall, who noted that bulls were encouraged to cover while still yearlings, which was still unusual in most of the rest of the country, and that nurse cows were often used, with young bulls remaining on a milk diet for between six and twelve months.[98]

Plate 15. The *Lincolnshire Ox*. Another early specimen of the improving Shorthorn breed. This animal was bred by John Cribbens of Long Sutton in Lincolnshire in 1782 and brought to London and exhibited for its remarkable size in 1790. A painting was made of it by George Stubbs in 1791 and engraved by his son, George Townly Stubbs. This print was published in January 1798. Photograph from the Museum of English Rural Life, University of Reading.

This feature can only be adjudged a novelty by the standards of arable husbandry. In those marginal, pastoral regions too far away from a commercial market for milk products, calves undoubtedly ran for a long time with their dams as a matter of course (for instance Highland cattle in the seventeeth and eighteenth centuries). The Lancastrian system probably originated as a persistent element of this essentially 'unmanaged' and 'wild-type' behaviour, for bull-raising only, while female calves and those destined to be castrated were quickly removed to allow the milk or milk products to be sold. The procedure was novel in arable and grazing regions near markets for milk and beef because commercial pressures and lack of land had long since prevented the practice of running any calves on in this way. It seems that the properties of early maturity and good beef conformation (in eighteenth-century terms) were more likely to persist among marginally managed, semi-wild cattle than among stock fitted into a managed, lowland, agrarian economy. The 'improvement' of cattle with respect to beef conformation and early maturity could be seen as essentially the 'recovery' of these characters which were originally present, but debased and bred out by the side effects of close management.

In Lancashire itself, at the end of the eighteenth century, little vestige of the fine cattle or their method of management seems to have remained. Campbell reported in 1793 that the Longhorn cattle of the county were in a very middling state. The supposedly best bull in the county was certainly tolerable, but Campbell found his calves to be poorly kept and half-starved. Similarly, in Westmorland and Craven, the stock breeders were supposed to have been tempted by high prices to part with their best stock to Midland breeders, so that nothing of note remained behind.[99]

The common bull and the possibility of improvement among small owners

So much of the surviving evidence about husbandry comes from the estate records of the larger gentry and from literature written with the gentleman owner in mind, that there is a danger of ignoring the potential for improvement which may have existed among smaller owners. Consider those farmers whose herds were so small that they used either a common parish bull, or in those regions where common agriculture was not the standard practice, employed the services of some local bull for a fee.

The presence of common bulls can be traced back a long way in the records of manorial courts, one of whose major functions from the fifteenth century onwards was the regulation of agricultural practice in open-field villages. The responsibility for finding and keeping the common bull sometimes fell on the lord of the manor (or his lessees if he was not cultivating demesne lands himself), sometimes on the parish incumbent, sometimes on the chief parish officers, and sometimes upon a committee drawn from among all the officers. There does not seem to have been a consistent pattern and the burden of ownership could change from one type of individual to another within the same village or township.

While the provision of the bull was a charge on a particular member of the community for which he could receive no recompense, the incentive to provide only the very cheapest bull that he could lay his hands on was very great, although the breeding structure of village herds was such that many more cows were inseminated by one bull than was likely to have been the case with the herds of private owners of bulls. The potential for selective breeding was therefore greater, although the realisation of any scheme of improvement would seem unlikely. In addition, it is quite possible that bulls were changed on a yearly basis, so the chances of any one sire influencing the structure of a breed or race to any great extent was negligible. While it may have been true that to hold land and stock outside the communal system did give a greater opportunity to experiment with breeding, the common bull system, if an enlightened parish or manorial lord had so wished, could have provided a better way of disseminating the genes from good stock through the parish, than had the practice of the private ownership of bulls. Certainly, in the late eighteenth century, philanthropic landowners concerned with improvement adopted the system of making a good bull available free to any tenant who desired its services, an administration exactly parallel to the earlier common bull provision by the lord of the manor.

There is evidence that in the North, where common-field agriculture had never been a feature of village organisation and the administration for the provision of a village bull did not exist, small cowkeepers used bulls owned by neigbouring yeomen and gentry. In some cases a small charge was evidently made for this service by the late seventeenth century. For instance, the accounts of Sarah Fell, wife of George Fox, the founder of Quakerism, at Swarthmoor Hall in the Furness region of Lancashire in the 1670s, show a large number of entries for 'cow-bulling' charged to her smaller neighbours at 6*d* a time.[100] The practice was evidently not confined to the North, since a charge of 6*d* was paid by a small farmer in Kent (another region outside the common-field area) for the use of a bull in June 1686.[101] The bull or bulls owned by Sarah Fell and her mother were evidently stall-kept since in December 1674 they paid workmen for 'paving the bull house' among other repairs. As with stallions at this period, the cows were evidently brought to the bull where he stood in a stall or yard, and a charge levied for each covering.

The development of this practice of renting the services of a bull obviously has implications for the origin of stock improvement schemes, since hiring-out improved males for service was a standard part of the methods of specialist eighteenth-century livestock breeders. If the users of stock could be persuaded of its superiority they might be prepared to pay more for its service and thus provide an incentive towards the development of even better stock, for which higher charges could be made. In the case of the Fell accounts it seems their activity was merely an extension of an old social obligation on the part of larger farmers towards smaller ones. Although they were not making any obvious profit from bull letting and were certainly not equipping themselves, judging from the prices paid, with the desirable young bulls for which the region was famous, they were evidently covering their costs. During 1674 the services of their bull produced an income of £2 8*s*, which was more than enough to cover the difference between the one bull purchase recorded (£3 14*s* 3*d*) and the one sale (£2 15*s*).

It could be argued that in the open-field villages the manorial court and its representative officials may just as well have been progressive as conservative. Within the framework of their administration, the rotations and methods of open-field agriculture definitely changed and evolved over time. The villagers would have needed to spend more than the bare minimum on a bull, but if some machinery had evolved for recovering the cost and the villagers believed the increased burden was justified by the improvement of their stock, there would have been no major impediment to change. After all, the costs of high-grade or imported stock bear far more heavily upon an individual than they do on a co-operative group who can share them. Such cost sharing would be the only way in which anyone but a substantial farmer or gentleman could take any part in the improvement of agricultural animals. The necessary conditions for an open-field village to take an interest in stock improvement would therefore seem to be (a) a method of recovering some of the cost of a good bull and (b) the purchase, upkeep and disposal of the bull passing into the hands of the villagers themselves and not remaining the responsibility of the incumbent or the manorial lord.

The running of collective agriculture seems to have moved firmly into the control of the tenants during the sixteenth and seventeenth centuries. The function of those officers who had agrarian duties was to administer the husbandry of the village, which was either framed or approved at the annual meeting of the community at the manorial court.[102] These burleymen, elsewhere called 'tithingmen' or 'fieldsmen', kept accounts and, where these have been preserved, they give the best picture of the operation of village agriculture. To say that the responsibility for the provision of the bull had passed definitely into the hands of the tenants themselves, there must be evidence that these fieldsmen or their equivalent were taking full charge of it. There is some evidence that this transfer of responsibility had occurred in some villages and machinery had evolved to allow for the costs of upkeep to be passed on to the users of the bull's services.

In some villages the responsibility still resided with the incumbent, or attempts were still being made to force him to take on this task.[103] At Purley in Berkshire the rector successfully opposed an attempt by the jury of the manor to force him to keep a common bull, since the Court Rolls did not record this as a custom of the parish.[104] It sounds as if the provision of a bull had passed to the villagers, but they were unwilling to bear the expense and were trying to pass it off in a traditional direction.

Elsewhere in Berkshire it had certainly passed to the parish, but was in the hands of non-agrarian officers, so the conditions for the group's self-interest to express itself in stock improvement were not fulfilled. The bull was seen as a necessary, but regrettable, expense. Thus, at Drayton, responsibility for the bull in the 1740s and 1750s resided with the Overseers of the Poor, clearly an example of a village equating the upkeep of the bull with the expenditure it would most like to avoid. The Expense and Receipt Book of the Overseers showed that they bought a bull or bulls regularly, between 1746 and 1761, for sums varying between £3 14s 6d and £1. Sometimes the bulls were sold, usually for less than was currently being paid for them (which suggests they were not fatted before sale), and occasional grazing and other upkeep costs were recorded.[105] Otherwise, the bull was presumably fattened and disposed of in the village at Christmas and its keep was found on the grazings allowable to the administrators and therefore cost them nothing. Although receipts were recorded in the book, there is no evidence that charges were being levied for the use of the bull. However, one interesting entry for 1747 was 'paid for a hierd bull 14s'.[106] This suggests that rather than go to the expense of buying another one, for they had already paid £3 14s 6d for one bull, the maximum price they ever paid, for some extra service they had rented one from a private landlord or farmer in the vicinity. This indicates that, apart from the use of a private bull for a small charge at the place where the bull was kept, it was possible to hire one for a period of time, presumably transferring him to some central location among the cows. This practice of renting the entire male out to a herd or by the season was to become a major money-raising scheme among Midland improvers in the later eighteenth century.

The correct hypothetical conditions for interest in stock improvement through the parish bull seem to have occurred at Bodicote in Oxfordshire, which was an open-field village until its enclosure in 1768.[107] Annual accounts were submitted by the constables, surveyors and fieldsmen, and the accounts of the latter show that they were responsible for the purchase and upkeep of the common bull. Unfortunately they seldom itemised their annual accounts, merely recording the sums received during the year, their total expenditure and the resulting credit or deficit for transfer to the following year's account. However, for three years they did render itemised accounts and there were significant entries. In May 1760, an ex-fieldsman, John Wilson, was contracted to supply the bulls for the coming year at a charge of £6 5s. The two other years show that they made some money from bulling fees. In both 1717 and 1720 the amount came to exactly £3 1s 8d, which suggests that each tenant paid a fixed customary fee for the use of the bull.[108] The administrative framework here was of the type which could have allowed the village farmers to decide collectively to go for an improvement in their livestock by the acquisition of better bulls.

Other small farmers who could not afford the trouble and expense of a bull and were outside communal village life continued the seventeenth-century practice of using the private bulls of their neighbours and were charged a fee accordingly. The procedure was widespread and a fee of 6d or 1s was standard throughout most of the eighteenth century.[109] Some farm and estate accounts allow analysis of the breeding policy adopted by their compilers. The general valuation of a bull on most farms was less than that of a good milk cow or heifer.[110] Some other aspects of management give hints that careful breeding of stock was not entirely neglected. Most of the accounts show that the herds of cows were mixed, composed of both bought-in and home-bred stock. Several show evidence of the adoption of the standard procedure for naming home-breds adopted by horse breeders from the sixteenth century: the use of the same name for dam and daughter, with 'old' or 'young' added to distinguish them.[111] The majority of bulls were young, usually two- or three-year-olds, as had become conventional by this date. They were usually fatted off and killed when they became sluggish at reproduction and dangerous to handle.

7

Breeding sheep: mutton displaces wool

Introduction

In medieval times sheep had been used to provide three basic commodities. Firstly, they had produced dung or tathe, transferring fertility acquired during daily grazing on the wastes and commons, which were too poor to cultivate, to the arable fields by nightly folding. Secondly, ewes were often milked and what evidence there is suggests that ewe's milk was an important product.[1] Thirdly, all sheep were clipped to provide fleece wool, in the thirteenth and early fourteenth centuries for export to Continental manufacturers, and from the mid-fourteenth century onwards increasingly for working up into broadcloth in the native English trade. The first two products supplied no commercial markets, but provided important elements in the infrastructure of the partly self-sufficient subsistence farming by which most of the population lived. Wool did supply a market, and this was sufficiently attractive even in the twelfth and thirteenth centuries for both lay and ecclesiastical landlords with large tracts of poor-to-marginal pastoral land to stock sheep and sell wool. With the reduction of population which followed the Black Death in the mid-fourteenth century and the continued buoyancy of the textile trade, sheep keeping remained a worthwhile proposition and the deliberate stocking of good land, as opposed to marginal pastures, began at this point.

From archaeological evidence of the sizes of their fleeces and skeletons, all medieval English sheep were small animals. Only a few flocks aspired to an average fleece weight of more than 2 lb. These fleeces were usually of a primitive structure, similar in general type to the present-day relict, short tailed breeds of the Orkney and Shetland Islands, either containing considerable quantities of hair, or showing a fibre diameter distribution wider and less specialised than is the case in modern commercial breeds with fine, short or long wool.[2] Most of this wool was probably of middling quality and used for the production of middle- and poor-quality cloth. However, some sheep, especially those in the western uplands in Herefordshire, Shropshire, Staffordshire and the Cotswolds, and initially on the Wolds in Lincolnshire and Yorkshire, produced fleeces with fine wool essential for the manufacture of high-grade cloth. The western regions retained this reputation well into the sixteenth century, but elsewhere the sheep were probably of a poorer quality to begin with and improved manufacturing and wool-sorting downgraded their value.

During the sixteenth and early seventeenth centuries there is little evidence that the size or fleece weight of the average sheep in England increased to any great extent over the size and weight of animals in the later Middle Ages. There were no extensive improvements in general husbandry in most of England which would have supported larger, more productive animals. A whole battery of new agricultural techniques became available in the early seventeenth century, and these were reasonably widely adopted in the late 1600s and early 1700s.[3] The higher yields associated with these procedures allowed both a greater number and a larger size of animal to be kept.

Even with improved husbandry, larger sheep would not have been bred unless a new market for sheep products had existed. This was probably provided by a rising demand for better quality mutton and lamb, which came from animals specifically raised and killed for that purpose rather than from worn-out and aged sheep. Larger sheep meant larger fleeces and, generally speaking, heavier wool was more profitable to most farmers. It may have commanded a lower price per unit weight than the finest shortwool but since the majority of sheep had always produced only moderate fleeces, for most keepers the increased weight and length did not actually mean that the price of their wool fell, while in some cases it actually rose. Most therefore benefited from increased wool profits as well as from the enhanced value of their carcasses. It is probably true that the differential size between sheep kept on pastures, compared with those maintained on the arable as field sheep, originated in the fifteenth century or before, but even in the sixteenth and seventeenth centuries there is little evidence that 'pasture' longwools had diverged from 'arable' shortwools to any great extent in size or fleece weight and quality. But as the seventeenth century progressed and the market for meat animals grew, the size of lowland and marsh sheep, able to respond physiologically and genetically to the better nutrition available, led to the production of heavy, long-legged, long-woolled sheep whose fleeces were suited only to a restricted part of the worsted manufacture and whose meat became fatty and insipid. The centres of this transition were the rich marsh grazings of Lincolnshire, Somerset and Kent, the two former benefiting from large-scale marsh drainage in the mid-seventeenth century. Sheep of the same general type occupied the eastern and southern lowland pastures,[4] and animals from these latter regions were imported on to upland farms in large numbers from the late seventeenth century so that these farmers too could take advantage of their greater production potential. The drive for size as a response to better markets and improved nutrition was undoubtedly naive compared with the sophisticated analysis of the major components in animal productivity provided by Bakewell in the eighteenth century, but against the background of the small, bony and primitive sheep of the medieval period, the large lowland and marsh breeds were more efficient animals.

The sheep meat market may be said to have been firmly established, with evidence of early fat-lamb production, in the seventeenth century. This coincided with the development of stratification in the sheep trade, parallel to that which occurred with cattle at the same time. The Welsh droving trade in sheep was well established by then and this can only have been to supply meat, since the Welsh fleece was so poor.[5] Western sheep from Wiltshire, themselves feeling the benefit at source of the floating of the watermeadows, were moving to the eastern grazing counties through Weyhill and other fairs in increasing numbers in the seventeenth century.[6] The western uplands were becoming breeding regions and the eastern lowlands feeding regions, a pattern that was to become fully developed in the early eighteenth century. This was partially reversed later by the introduction of rotations based upon the turnip in the light soil regions of Norfolk and the Home Counties and on the previously barren uplands of Gloucestershire, Lincolnshire and Yorkshire. The changes in sheep population in these regions and others between say 1680 and 1770 were dramatic and perhaps represented as great a revolution in the type of livestock kept, and of their productivity, as the much better understood changes that occurred after the widespread adoption of the New Leicester and the improved Southdown over much of the country during the traditional 'agricultural revolution' from 1770 to 1850. These two southern and eastern breeds were exported back into the western uplands as improved animals during this period, thus completing the reversal of the sheep stratification which had existed in the early eighteenth century.

Stasis and change among regional sheep types

Descriptions of sheep were extremely rare in the sixteenth and seventeenth centuries and no good general picture of the variant types can be constructed from contemporary sources. The only general description of the range of sheep in England was written by Gervaise Markham in 1631.[7] Other stray references to regional sheep can be found in seventeenth-century sources and occasionally in sixteenth-century ones as well.[8] Wool price schedules and other economic documents of the wool trade provide some circumstantial evidence about regional forms.[9] Only in the Midland counties, where most of England's wool and mutton were produced, is it sensible to speculate on potential breeds and breed changes.

According to Markham, all the sheep of the heartlands of the Midland Plain were of the same general form – well-shaped, large and of the deepest staple – with the sheep of the Lincolnshire marshes being of the same type but significantly larger. The Midland sheep were not all uniform. Mid-sixteenth-century price lists indicate that the wool of the West Midlands was superior to that of the East Midlands in quality, a divergence which seems to have occurred between 1450 and 1500. This downgrading of East Midland wool may have been produced by the spread of incipient longwools on the Lincolnshire marshes into the surrounding eastern counties, where interbreeding would have lowered the quality of the fleece somewhat. In addition, writers on wool in the sixteenth and seventeenth centuries usually distinguished between two sorts of lowland sheep, referring to an 'arable' or 'field' sheep as opposed to a 'pasture' type.

In 1547, the term 'pasture' sheep was used to account for the greater wool output in the time of Henry VIII compared with that in the reign of Edward III, with the obvious implication that such pasture animals were bearers of heavier fleeces than the normal sheep of the lowlands. It was estimated that 20 fleeces had made a tod in Edward's time, but that 15 was the average now (1547), the difference being ascribed to the greater incidence of heavier pasture fleeces, which were perceived as being of relatively recent origin. The two Midland types were not contrasted in any obvious way by morphology or fancy points and it is possible that they were derived from a common ancestor and diverged as they came to occupy increasingly different roles in the fifteenth- and sixteenth-century Midland agrarian economy. Before the advent of convertible husbandry in the region in the seventeenth century, there was sharp divergence between sheep walk and arable villages. As will be seen, some Midland sheep graziers on the permanent grassland were definitely 'improvers' and may have hastened a spontaneous divergence of type through differential selection under different management conditions. Alternatively, the pasture sheep may have had a separate origin from arable types. A possible source from the Lincolnshire Wolds will be suggested below.[10]

These sparse early data on sheep varieties could not be properly elaborated until the end of the eighteenth century, when comprehensive descriptions of local sheep breeds began to appear. By this date at least thirty or more distinct breeds were recognised. Description and classification of these sheep present problems because they were undergoing transition because of extensive cross-breeding and breed replacement. By this date

Bakewell's New Leicester had grown from a local craze to an almost universal fashion and local breeds had been greatly obscured by interbreeding with it. When the second editions of the *Board of Agriculture County Reports* were published in the first two decades of the nineteenth century the influence of the New Leicester had been compounded by the use of John Ellman's Southdown and the import of Spanish Merino finewool.

Several lists and distribution maps of these breeds have appeared in the literature. Trow-Smith has given the fullest detailed descriptions of the breeds, drawn from contemporary sources.[11] Ryder has grouped the breeds in classes based upon biological criteria and has constructed a map of the distribution of these classes.[12] Other accounts have classified breeds according to their role in the agrarian economy or by the type of fleece wool they supplied to the textile industry.[13] All these reviews differ both in detail and principle as the purposes of classification are different in each case.

It is possible, by reading rather carefully between the lines of the first editions of the *Board of Agriculture County Reports*, supplemented by other references to sheep which occur elsewhere in the literature, to construct a speculative picture of the distribution of breeds over England at the end of the seventeenth century and to see how this differed from that found at the dawn of proper journalistic accounts of the breeds in about 1770. The changes may be explained to a very large extent by the adoption of improved systems of crop husbandry during the early eighteenth century and perhaps in some cases by the application of selective breeding techniques.

Throughout the period it remained possible to distinguish four basic classes of animal according to their function in the agrarian economy. (1) Sheep of the forests and heaths, which spent most of their time on barren sheep walks and which may or may not have had a role in the fertilisation of small amounts of arable land by a system of folding. (2) Down, fallow or common sheep, whose main sources of food were the stubbles and fallows of the lowland villages practising unimproved agriculture. Their main function was to fold the arable and they spent only the minority of their time on common sheep walks. These were the 'field' or 'fallow' sheep of the Midland counties and may or may not have been essentially the same type as those above. (3) The larger 'pasture' or 'marsh' sheep, kept solely as grazing sheep and showing little inclination to travel, search for their food or stand close folding on arable or stubble. (4) Mountain sheep of the exposed and barren uplands, living essentially wild on moors and the higher hills and mountains. They may at times have done some folding but their primary virtue was their ability to survive hard conditions.

These categories are utilitarian and not biological. What follows is an attempt to group the various sheep types by broadly biological criteria in the light of the contemporary literary descriptions. These groups show strong

Breeding sheep

Plate 16. The Old Wiltshire breed. A lithograph from a painting by W. Shiels executed for the second volume of David Low, *The breeds of the domestic animals of the British Isles* (London, 1842). By the mid-nineteenth century the Wiltshire breed had been considerably modified from the rangy, upland and poor-soil sheep which had been successful in the mid-eighteenth century.

parallels to Ryder's classes but differ in detail, and on occasion fundamentally, from his. The differences reflect Ryder's use of his extensive first-hand knowledge of sheep biology and archaeology as the basis for his classes, whereas the groups employed here draw their biological basis only at second-hand from Ryder (and other sheep biologists and enthusiasts) and depend largely on historical, literary evidence.

It must be stressed again that the first description of sheep types in the late seventeenth century is highly speculative and that the number of varieties recognised with precise morphology in both this description and that of the late eighteenth century are probably too simple. Evidence from cattle herds showed how widespread were colour and marking variation even in areas where the literary accounts suggested uniformity. The breeds here are based almost entirely on literary descriptions and must therefore be similarly suspect. It may be that the concept of a 'breed' in the twentieth-century sense of a group of domestic animals sharing a large number of common morphological features by virtue of genetic homogeneity, is wholly inapplicable to the regional forms outlined here. They may merely have represented the stock found in a particular geographical region and not have been particularly homogeneous. Even when modern relict breeds under primitive management remain isolated, the management and selection pressures working on them seem to favour the survival of diverse morphologies rather than tending towards similarity of appearance.[14] First of all a classification is given for the late seventeenth and early eighteenth centuries.

Group 1
Largely unpigmented, hornless sheep, often with a top-knot of wool on the forehead. Fleece types various: fine-, medium- or long-woolled. All fleeces with a high reputation for whiteness
A historical case can be made for supposing that three of the four groups here had a common biological origin, while the first was probably independent and of a different basic type. The least highly 'evolved' of these breeds were often reported to have had pigmentation of the face in the early eighteenth century.

(a) Sheep of the South Welsh Marches (1a on Chart 1)
These were known in the eighteenth century as Hereford or Ryeland sheep and were direct descendants of the famous finewools of Leominster in the Middle Ages. At the end of the seventeenth century they were probably still the most abundant breed of sheep in most of Herefordshire and Monmouthshire and in a narrow, northwards stretching belt through western Shropshire, eastern Radnor and eastern Montgomery. They were evidently of the heath or forest type since they were never used for folding on the arable. They were sheltered in cotes on the slopes of the hills in winter as they were too delicate to stand penning in the open. They were, however, very tolerant of near-starvation and could survive on little food and search tirelessly for it. It is unlikely that these sheep were related biologically to the other three classes, firstly because they were not fallow folding sheep or pasture sheep at any stage of their development, secondly because the archaeological evidence from the Middle Ages suggests that the fleece type of sheep producing the very finest wool was different in structure to any other fleece type of the period, and thirdly because the continual nineteenth-century attempts to 'improve' the Hereford by crossing with the New Leicester proved unsuccessful and showed these two sheep to be of essentially incompatible type. By contrast, the intercrossing of Leicester and Lincoln sheep with those of the other classes included here were successful and suggest a basic similarity of type.[15]

(b) The field sheep of the Gloucestershire and Lincolnshire uplands (1b on Chart 1)

The fleeces from sheep on the Lindsey uplands of Lincolnshire and the Cotswolds of Gloucestershire had been the second-rank growths of the Middle Ages, although during the fifteenth and sixteenth centuries the Lincoln wool seemed to deteriorate by comparison with that from the Cotswolds. In the seventeenth century the sheep of both these regions were grazed for most of the time upon the barren sheep walks of the hilltops and probably spent some of their time folded on the arable corn lands in a classic upland, light-soil economy. It is possible that the sheep of the associated East Yorkshire Wolds were of the same general type, since the Lincolnshire and Yorkshire Wolds are contiguous and the agrarian economies of the two regions were very similar. Henry Best hinted that the sheep he was keeping on the Yorkshire Wolds were a very mixed group, with both polled and horned animals and white- and black-faced variants. Since there does not seem to be an extant description of the face colour or horn characters of the Cotswold or Lindsey animals in the seventeenth century it is possible that these populations too were mixed with respect to these characters. On the lower slopes of both the Lincolnshire and Yorkshire Wolds, larger sheep of the pasture type were kept in the seventeenth century, a practice which may have arisen from selection among the upland sheep on the better grazings of the lower slopes. These pasture sheep of the eastern, seaboard counties were part of the third group in this class, the long-woolled, lowland, grazing sheep.[16]

(c) The lowland pasture longwools (1c on Chart 1)

The pasture sheep of the Midland counties and the marsh sheep of Lincolnshire already had a deeper fleece than the local field sheep in the early seventeenth century. The difference in size and fleece yield grew as the century progressed and Midland pasture quality improved with the introduction of up-and-down husbandry and the use of artificial grasses. By the end of the seventeenth century it would seem, from the sparse fleece weight data that has survived, that the largest pasture sheep were yielding fleeces of about 4 lb to 5 lb in weight. By the opening of the eighteenth century sheep of this heavy-fleece type were grazing most of the suitable pastures of Lincolnshire, Yorkshire, Durham, Northumberland, Leicestershire, Rutland, Northamptonshire, Cambridgeshire, Warwickshire, Oxfordshire, Somerset, Devon and the east of Kent. The smaller field sheep were still present in these counties but the longwools may be said to have been the most obvious feature of the whole region. They were also present in the better pastures of the North and West Midlands, but since there were more heaths, wastes and forests there the longwools were not so prominent in the

Plate 17. The Old Norfolk breed. Lithograph from David Low (1842). Even the 'old' breed of the mid-nineteenth century must have been severely modified from its successful mid-eighteenth-century predecessor. The hornless lamb in the background is a cross with the Southdown, the breed which displaced the Norfolks and the Wiltshires in the early nineteenth century.

sheep husbandry of those regions. It is a matter of conjecture at the moment as to whether the pasture longwools evolved from the local field sheep of their own regions or whether there was a centre where the basic longwool type developed and from which it spread into the improving pasture lands of the seventeenth and eighteenth centuries. The wide geographical distribution of the longwool type might be said to favour the former possibility, while its universal morphological features of white colour, fleece type and hornlessness might be said to argue for diffusion from a local centre of evolution, given the wide variation in the superficial morphology of many of the breeds of heath and fallow sheep in the zones where the longwool dominated. If a single source for their development is sought the most probable candidate is the old Wold sheep of High Lindsey. At the end of the seventeenth century the largest examples of this longwool breed were the Lincoln of the marshy coast from Boston to the Humber, and the Teeswater in the Tees valley. In the second rank were the sheep of Leicestershire and Warwickshire.[17] In the more peripheral regions there were, at the end of the eighteenth century, variant forms with coloured legs and, occasionally, horns. This might be taken as evidence for a local origin of these breeds from pigmented heath, down and fallow sheep, but the contemporary reporters themselves felt it was more likely to be the result of a relatively recent crossing of longwools with the still extant heath types of these regions.[18]

(d) The Cheviot breed of the Northumberland hills (1d on Chart 1)
Ryder suggests that this breed originated from an ancient Scottish border tan-faced, horned type and that it has no relation to the Lincolnshire Wold sheep or the longwools. However, most eighteenth-century reports speak of a breed that was white-faced and polled in both sexes, although there had been a considerable amount of facial pigmentation early in the century, which had not entirely disappeared a hundred years later. These contemporary authors believed the dun-faced Grampian of the next range of northern hills to be an entirely different breed. An origin from a horned, mountain breed also seems unlikely given the later history of the Cheviot breed. In the mid-eighteenth century, improvement of the Cheviot breed was successfully begun by crossing it with Lincoln rams, a type of cross which seldom proved successful when a longwool was crossed with a mountain breed. There are also those who claim that the nineteenth-century variant of the New Leicester, today known as the Border Leicester, contains a considerable amount of Cheviot ancestry, again indicative of compatibility between this type and the lowland longwool. The green-hill Cheviot may have been an animal very like its heathy cousins on the Yorkshire and Lincolnshire Wolds in the seventeenth century and probably functioned in the same way in an unimproved upland economy.[19]

Group 2
Southern and western mountain breeds with tan faces and horns in both sexes

(a) 'Celtic' sheep (2a on Chart 1)
Very small, horned, dun-faced sheep occupied the whole of the mountain regions of Wales and southern Scotland that were not given over to cattle breeding. These sheep were not especially hardy when compared with black-faced heaths, Herdwicks or modern Cheviots and were confined to the more sheltered valleys. Sheep of very similar general character – such as the Exmoor, Dartmoor and Mendip Hill sheep – occupied the upland regions of south-western England. An eastern extension of this type – such as the Dorset Horn and the rare Portland breeds – occupied the downlands of Dorset and the surrounding regions.[20] The Dorset Horn was famous in the eighteenth century and afterwards for its exceptionally long breeding season, so that winter and spring lamb could be raised from the housed ewes as 'Christmas' or 'house' lamb. To some, the long breeding season has

suggested a separate origin for this breed from somewhere to the south of Great Britain.

(b) The Wiltshire Horn (2b on Chart 1)
On the downs of Wiltshire and Hampshire a larger breed of this south-western type was found. This was the Wiltshire Horn, a tall, very tough sheep, ideal for folding. Late eighteenth-century writers generally thought that the differences between this type and the neighbouring Dorset were sufficiently great for a number of them to question any common origin.[21] It also resembled two adjacent breeds that fall into different categories on the classification employed here: the old Hertfordshire and the Berkshire Knott.[22] It is possible that these latter breeds, on the boundary between the Midland Plain, increasingly dominated by longwools, and the small Dorset, Sussex and Surrey Down breeds, represented transitional types stabilised from consistent crossing.

Group 3
Small and middle-sized black-faced and horned heath and fallow sheep. These were either fine- or middle-woolled
Breeds of this type were found in two distinct geographical areas in the seventeenth and eighteenth centuries. They may originally have been connected across the Midland Plain.

(a) The Black Faces of East Anglia (3a on Chart 1)
These were all folding, fallow sheep grazed upon the arable stubbles and fallows and on the barren sheep walks of the poorer soils of Norfolk, Suffolk and the north-eastern Home Counties. The most famous representative of this type was the Norfolk Horn, noted for its ability to get a living on the unimproved wastes of Norfolk, Suffolk and Essex, and in the London meat market for its flavour. In the other counties of the area, less favoured variants occurred as the native breeds of Bedfordshire, Huntingdonshire, South Cambridgeshire and Hertfordshire. The Hertfordshire was an ugly, large breed and in some ways it resembled the Wiltshire Horn, which may have played some part in its ancestry.[23]

(b) The Black Faces of the western commons (3b on Chart 1)
Markham described a black-faced finewool of this region, and the heath and forest sheep of Shropshire, western Staffordshire and western Worcester-shire in the late seventeenth and eighteenth centuries were their linear descendants. Outliers of this type could also be found to the east. There were certainly sheep of this general form on Clifton Heath in northern Leicestershire.[24] Some of the sheep that remained on the commons and

Plate 18. The Romney Marsh breed. Lithograph from David Low (1842). The ancestors of this breed were the Kentish branch of the long-woolled sheep type of the seventeenth and eighteenth centuries.

poor-agriculture regions of Warwickshire[25] were also of this type and it may well be that their distribution was widespread in the early eighteenth century, scattered among the dominant longwools.

Group 4
Small and middle-sized grey, brown or speckled-faced, polled down or fallow sheep. These were either fine- or middle-woolled
This group also tended to fall into two geographical regions in the late seventeenth century, one in the South-East and one in the North and West Midlands, although it would seem likely that there had once been extensive connections through the Midland counties on a north–south axis.

(a) The polled Black and Brown Faces of southern England (4a on Chart 1)
This group flourished on the chalk downs of Surrey and East Sussex and the barren heaths of West Surrey and Hampshire. The best-known variants were the fine-woolled sheep of the South Downs, from the river Adur eastwards, and the Banstead Down finewools found on the North Downs.[26]

An apparent extension of this type occurred in the polled Berkshire Knott breed of the downs of that county, but this may have been another transitional breed between the southern finewools, the Midland longwools and the western, horned types.

(b) The polled common and forest sheep of the North and West Midlands (4b on Chart 1)
This was another group which appears to have been a linear descendant of the type described in the area by Markham in the early seventeenth century. Late in the seventeenth century they were still common on the heaths, forests and poor arable regions of Worcestershire, Staffordshire, southern Derbyshire and Nottinghamshire, counties in which the penetration of the pasture longwool sheep was by no means complete by this date.

Group 5
The white-faced, horned, close-woolled breeds of the western Pennines (5 on Chart 1)
The surviving representatives of this type in the late eighteenth century were the white-faced Woodland and Penistone breeds of the Peak and southern Pennines and the Silverdale of Westmorland and north-west Lancashire. These islands of sheep of essentially similar type probably represented the extreme ends of a belt of such sheep that had occupied the western Pennines from the Peak to Westmorland in the seventeenth century.

Group 6
The black-faced, horned, coarsewools of the northern mountains and moorlands (6 on Chart 1)
These were pure mountain sheep found in the Pennines, Peak District, Lake District and North York Moors in the late seventeenth century. By this date they had begun their penetration of the Scottish Highlands but had probably spread little further than the Grampians, which were still occupied by the old Grampian Dun Faces. The most general type was perhaps similar to that now associated with the mountains north of the border, such as the Linton or Scottish Black Face. The origin of this variety is obscure. Superficially, they resemble the southern, middle-woolled Black Faces such as the Norfolk Horn, but the characteristic hairy fleece and tolerance to harsh mountain conditions rather argue against this, unless the selection applied to a common type in the two very different environments was sufficient to produce this divergence during the medieval and early modern periods. It is certainly true that the eighteenth-century Norfolk fleece was uncommonly full of hairs.[27]

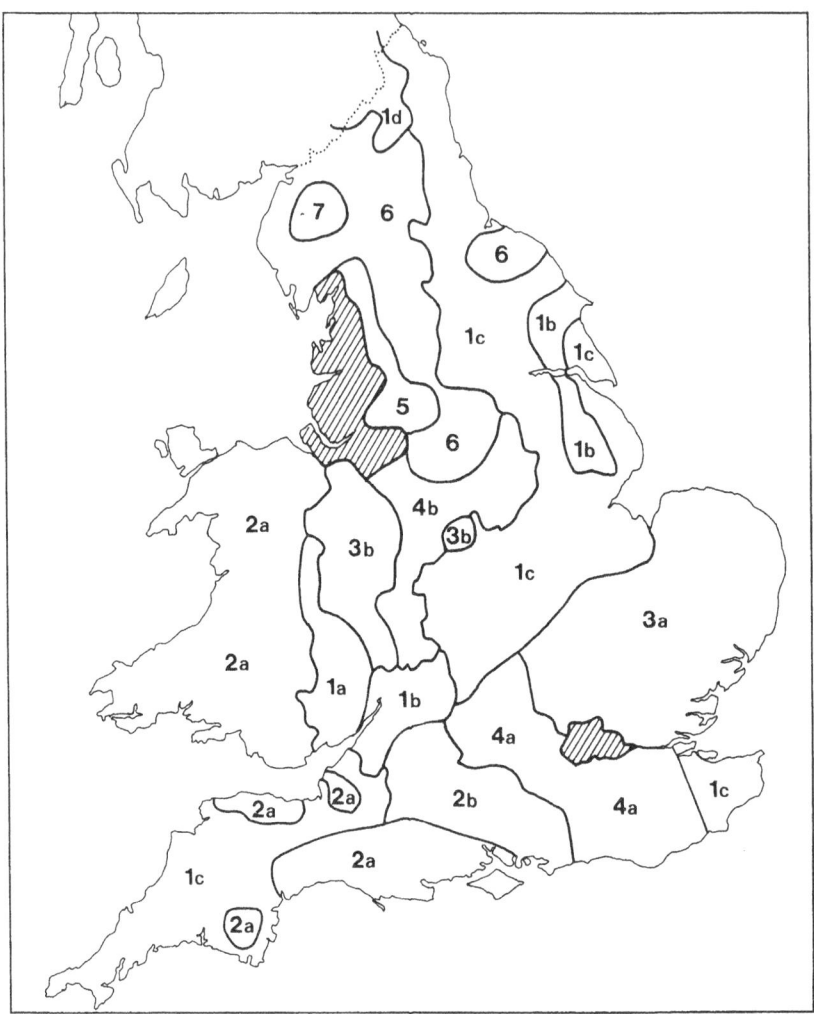

Chart 1. Approximate distribution of the dominant types of sheep in England and Wales in the late seventeenth century

Group 7
The Herdwick sheep (7 on Chart 1)
This was another small and very hardy true mountain breed, found on the most exposed portions of the mountains of the Lake District. The type seems originally to have been polled, with a speckled face, but in the modern breed the rams are horned, suggesting perhaps the influence of some crossing with the Group 6 Black Faces.

By the late eighteenth century the pattern of breeds and types of sheep across the country had dramatically altered. Most of these changes had already been set in motion by 1770 although much of the transition was not complete until the end of the century. The situation after 1770 was clouded by the spread of the New Leicester, and from 1790 by the development of the Southdown, but the presence of these two breeds did not alter the pattern of change which was already well under way. They merely provided better versions of the two types of sheep that were now beginning to dominate the whole of English agriculture. One type was the large, grazing pasture sheep of the lowlands, while the other was a middle-sized, fine-woolled, folding sheep of the more sheltered southern and Midland uplands. The development of this latter type was an eighteenth-century phenomenon as the improved husbandry of turnips and artificial grasses spread over these light-soil areas and made them more productive. The New Leicester was designed to fill the lowland niche in the late eighteenth century and the Southdown the upland, folding niche. But other breeds had already spread to satisfy this demand for more productive animals, long before Bakewell and Ellman began to improve their respective breeds.

The most obvious changes in the distribution of sheep during the eighteenth century occurred where improvements in husbandry were most marked. This was on the rich grazing lands of Lincolnshire, where there was drainage and reclamation of both salt marshes and inland fens, and on the light, but potentially productive, upland soils of the downs, wolds and hills of southern and eastern England. There was also improvement produced by enclosure and the laying-down of permanent pasture on the clays of the Midland Plain, which produced an increase in the amount of longwool grown on the backs of large sheep and continued a process of change which had been going on there since the sixteenth century. In nearly every case the response was a simple increase in the size of the animal, continuing the trend in animal breeding which had been common in the late seventeenth century. This was produced either by the selection of larger sheep within the local population, or by the introduction of larger sheep into a region from elsewhere, or by the 'grading-up' of stock through the introduction of rams from areas where larger sheep were raised.

The changes seem to have been successful and the increased productivity of sheep-grazing regions had several consequences for the structure of the sheep industry. Within each local region the improved productivity of the uplands or heaths broke the traditional pattern of the poor-soil farmers operating as stock breeders while those of the surrounding lowlands or better soils specialised in grazing and did little breeding of their own, merely buying in excess animals from the upland breeding economy. With better productivity the poor-soil breeders now had far more food upon which to keep their stock and could finish the sheep off for market themselves. There were fewer excess animals for sale to the valleys and the better-soil graziers were forced to start breeding stock for themselves. The more general, inter-regional effect was to increase the output of sheep from those parts of the country where sheep husbandry formed the backbone of the economy and to accelerate the trend towards regional stratification in sheep production similar to that which has already been noticed in cattle and horse husbandry.[28] Some regions, especially those around London, simply gave up sheep breeding altogether and merely bought in stock from other regions. To look at some of the details of this process of breed replacement and change, consider the distribution of breeds over the country as a whole in about 1770 compared with that of the late seventeenth century (see Charts 1 and 2).

Group 1

In its traditional regions in Hereford and West Worcestershire, the Ryeland breed was just about recognisable as the fine-woolled common and heath sheep it had once been. In 1791 this older sort of Ryeland could only be found on more remote farms and by the end of the century the average fleece weight in the area had crept up to 2 or 3 lb and there were even some weighing 5 lb.[29] The agricultural economy to which the old breed had belonged was fast disappearing and it declined into a sorry half-bred cross with the New Leicester in the first half of the nineteenth century. Only when the Shropshire Down arrived in the region in about 1850 did it become obvious how the Ryeland should have been improved. Its descendant today is to all intents and purposes a middle-woolled down breed, very similar to the other regional variants of the down group.[30]

Perhaps the most dramatic, but largely unrecorded change in this group, was the transition of Cotswold hill sheep from a heath variety of walks and arable folds in the old husbandry to a pasture longwool in the new. Whereas in other upland regions the change was often from a heath sheep into a larger and more productive variant, the Cotswold change was to a completely pasture form, where no folding whatsoever was practised as part of the new system. This change seems to have started early, for although Defoe still referred to Cotswold wool as second rank in the early eighteenth century, local users in the Gloucester finewool textile trade had already relegated the growth of the hills to the manufacture of second-grade cloths in 1680 and were using only Hereford and Spanish wool in their finest material. Throughout the eighteenth century the sheep of the hills were gradually changed by the continued introduction of long-woolled rams from the lowland Midland Plain in Warwickshire. By the time Marshall described the sheep in the late eighteenth century they were the same as the Midland longwools. He described the classic role that the hills had previously played in the economy as breeding zones for stocking the surrounding vales, but the hills were by then completely converted to turnip husbandry and artificial grasses and the farmers were able to get all their sheep into condition for the market on their own grounds.[31]

The same arable improvements and breed transitions occurred on the eastern Wolds of both Lincolnshire and Yorkshire, but the change was neither so dramatic nor so early. By the 1790s extensive enclosure and the spread of turnip and artificial grass husbandry had occurred on both sets of Wolds and the heath sheep of the uplands were rapidly being converted into longwools of the standard lowland type. By the early nineteenth century the sheep on the Lincolnshire Wolds were only slightly smaller than their marshland relatives, and the practice of folding had completely died out. What state this change had reached by 1770 it is impossible to say, although turnips did not appear in the region until that date. The change on the Yorkshire Wolds was even later than in Lincolnshire and, although the Wold farmers were much more productive and fatted their own sheep by the end of the century, the sheep of this region still did not have a fleece suitable for combing and worsted manufacture, even in the first decade of the nineteenth century.[32]

Long-woolled sheep had continued to spread into the better pastures of the Midland counties throughout the century, so that by 1770 the great majority of sheep on the whole Midland Plain were longwools and the heath and common sheep were confined to their barren, unimprovable sheep walks.[33] Although now firmly within the Midland longwool zone, Worcestershire, Staffordshire, Nottinghamshire and Shropshire still produced a very mixed bag of wools in the late eighteenth century.[34] Long-woolled breeds certainly extended their geographical range between 1700 and 1770 but whether they got any larger is open to doubt, although in the lowland zones of Lincolnshire and North Yorkshire the sheep do seem to have increased in size and yield of wool. This change was probably caused by deliberate policies on the part of breeders since the grazing was already so rich that there was little naturally selected or accidental improvement that could have occurred. Even in Lincolnshire not all the lowland growth was good combing wool. There were commons and heaths dotted over the county where wool growth was poor, and the same applied to the unreclaimed fens and bogs where only scruffy, half-wild sheep could be kept.[35]

The polled, hill Cheviots were a predominantly white-faced breed by the late eighteenth century and from about 1760 onwards improvement in the size and conformation of the breed had been sought by crossing it with lowland longwools, with considerable success. In the first instance this cross had been with rams obtained in Lincolnshire, although later in the century

Plate 19. The Old Lincoln breed. Lithograph from David Low (1842). The enormous fleeces of the eighteenth-century Lincoln longwools had been moderated by nineteenth-century breeders and more attention paid to carcass characters. Low here illustrated what he took to be a relatively unimproved (in nineteenth-century terms) specimen.

part-bred Leicester tups were used.[36] The influence of longwool blood did nothing to assist the fleece quality of this breed. In the unimproved state it grew a finewool fleece of about Southdown or old Cotswold quality. In the late eighteenth century the fleece was certainly not all of fine grade and this presented problems to the sheep improvers of Scotland, who felt their northern hills were most suited to the growth of short, fine-woolled fleeces.[37] Despite disappointment over the quality of its fleece the Cheviot breed was grown over much of the Highlands of Scotland in the first half of the nineteenth century.[38]

Group 2
Throughout the eighteenth century the export of both Dorset and Wiltshire Horn sheep from their West Country breeding grounds continued to expand. The Dorset continued to be used for the production of 'house' lamb in the counties around London and it was also imported into Kent and other Home Counties as a producer of early lamb on turnips or artificial grass. The majority of these West Country sheep were Wiltshire Horns, hardy enough

for lowland folding on the arable in winter, and to search for their food on the common sheep walks if this was necessary, but large enough to produce a reasonable carcass of mutton.[39]

The Wiltshire retained the short fleece characteristic of field sheep, and never grew combing wool or proved amenable to crossing with pasture longwools. Although slow maturing, they were ideal sheep for those areas where soils were too poor to become productive under the new husbandry and they spread out from the sheep–corn country of Wiltshire and Hampshire across the uplands of Berkshire and the Chiltern counties, into the heaths and downs of Surrey and Kent and the poorer-soil regions of Hertfordshire, Cambridgeshire and Essex.[40] The native breeds of Berkshire, the North Downs and Hertfordshire disappeared before the onslaught. Although the fleece weight of the Wiltshire never seemed to average more than 3 lb, its quality certainly deteriorated during the seventeenth century, probably as a result of changes in fleece texture and type as it fed on the improving pasture of its native watered meadows. The same deterioration in the quality of Hampshire wool was also observed in the 1690s, perhaps as a result of the spread of the enlarged breed of Wiltshire into that county.[41] Lisle commented with disfavour that the sheep from Wiltshire did not do well on the exposed and cold fields at Crux Easton in Hampshire at the opening of the eighteenth century, possibly because they were already improved to the extent that they could not tolerate a really bare living without the better winter feed provided by the early spring bite of watered meadows.[42] It was reported at the end of the century that they had recently been enlarged by crossing with bigger rams and that they used not to be so rangy and tall as they were by that date.[43] However, it seems just as likely that the differences in size evolved on their native pastures during the eighteenth century.

Group 3

Exactly the same general phenomenon may be observed in East Anglia and the eastern Home Counties as was seen in Group 2 and the upland regions of Group 1. The Norfolk sheep of the late eighteenth century was a larger breed than it had been before the advent of improved light-soil husbandry. Marshall noticed that the older form still survived on the more barren heaths of the county and that they were identical in appearance to the general sheep of the region, except that they were much smaller and had finer wool. The size of the new eighteenth-century Norfolk, which was found throughout Suffolk as well, was about the same as that of the Wiltshire Horn. The two breeds had evolved in parallel in response to an identical set of conditions. These larger Norfolks shared with the Wiltshires the grazing pastures of Essex, Hertfordshire and the poorer soils of Bedfordshire, Buckinghamshire and Huntingdonshire. The older, smaller, local black-

faced varieties of these counties had all but disappeared by the end of the
century.[44] In the North-West Midlands the other group of breeds of this type
were now to be found as breeding flocks only on the forests and commons of
Shropshire and some parts of Staffordshire and Leicestershire. Little was
done to improve or alter these sheep because their grounds were too poor for
effective new husbandry.[45] Like the Herefords, their development as a down-
type middlewool had to wait until the mid-nineteenth century when
Southdowns penetrated into the region in force and crosses with this type
gave rise to the modern Shropshire Down and the cluster of local variants in
its immediate vicinity which were based on Welsh as well as Shropshire
common sheep.

Group 4

Sheep of this type retreated before the spread of larger breeds during the
eighteenth century. The small, hornless sheep of Surrey and North Kent
virtually disappeared, while the Berkshire Knott became a small enclave of
hard-pressed sheep on their native downs. In the North Midlands these
heath and forest sheep were only found on their barren heaths at the end of
the century, as pasture animals took over on the permanent grassland that
was spreading across the Midland clays, while the arable farmers had
recourse to the Shropshire hills for stock because the local heath sheep were
too poor for lowland farmers to consider grazing them.[46] The only area
where a breed of this group survived unscathed was on the South Downs.
The western, horned breeds never seem to have penetrated into Sussex
further than the River Adur. The Southdown evidently had virtues greater
than mere local prejudice and its isolation from the rest of South-East
England behind the impenetrable Weald. After the demise of the Cotswold,
higher Lincoln and Wiltshire wools, it remained the only high-grade
finewool in the country apart from the small quantity of Hereford wool. Its
fleece did not deteriorate badly as the husbandry of the Downs became more
productive, and its mutton quality and conformation were good enough for
John Ellman and others to create an improved breed that would itself
displace both the Wiltshire and the Norfolk from their pre-eminence as
folding, light-soil sheep over all the areas that the two latter breeds had
conquered during the century. An experimental foray was being attempted
into West Sussex with Southdowns in the early 1770s, and by the 1790s the
boundary between the western Horns and the Southdowns had moved
down the coast from the Adur to the Arun.[47] By the early nineteenth
century the Southdown was being widely introduced in the native county of
the Wiltshire Horn, and in Norfolk and Suffolk was being crossed with the
larger Norfolk breed to produce the Suffolk Down breed, which would
entirely displace the Norfolk from its own county as well.[48]

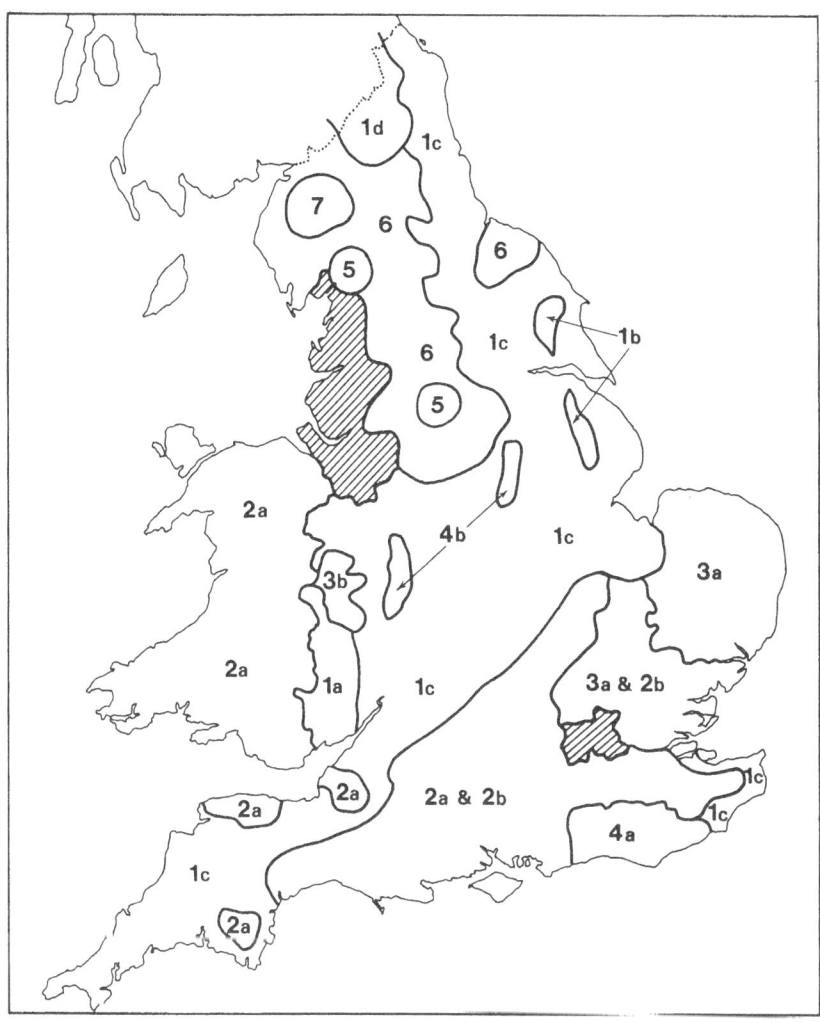

Chart 2. Approximate distribution of the dominant types of sheep in England and Wales, *c.* 1770

Group 5

Whatever the original extent of these close-woolled, white-faced, horned sheep, they were evidently in retreat during the eighteenth century before the advance of the hardier Black Face. By the end of the century the Penistone, white-faced Woodland and Silverdale were already to all intents and purposes relict breeds which would play no further role in the mainstream of sheep breed development.[49]

Group 6

In England the general status of the Black Faces did not appear to have changed, although this may merely reflect an almost complete absence of records. As in the case of the Cheviot, the main thrust of Black Face development was north of the border in Scotland, where the two breeds disputed territory throughout the whole of the first half of the nineteenth century. This penetration of true mountain and hill sheep into Scotland was a phenomenon of the late eighteenth and nineteenth centuries since the earlier, pre-clearance Highland economy had been based to a large extent upon cattle breeding and subsistence agriculture.[50]

Group 7

The territory of the Herdwicks did not change to any extent during this period, and nor has it subsequently. As a general mountain breed it has always been too small to compete on a wide scale with the Black Face.[51]

Quantitative evidence of carcass weight and fleece yield

The largest group of abstracted and published carcass weight records from the seventeenth century are those given by Thorold Rogers of sheep killed for the table in Winchester College between 1644 and 1699. As with the oxen recorded in the same source, the basis upon which the data were prepared was not given, so it is impossible to say whether the annual values given by Rogers represented the average of several purchases, a single selected representative value, or a single annual purchase for some specific occasion. One must assume that the weights recorded were those of fat adult sheep, rather than lean or young animals and that the weights were the killed-out weights of edible carcasses. It is not safe to assume that the source of animals was either local or consistent since Rogers noted that the bursars of some Oxford colleges and of Eton College travelled some distance to secure their supplies. From his comparison with less comprehensive data from other sources Rogers did not believe the Winchester weights were representative as they were lower than those obtained elsewhere.[52]

Apart from a few isolated years, the recorded weights seem remarkably consistent from year to year compared with other estate data given below. The mean weight for the series was 40.32 lb, which, on the basis that killed-out weight was 60% of live weight, meant the sheep themselves could not have weighed more than 70 lb. Among modern breeds, only the Welsh Mountain sheep, the relict Portland breed and the primitive breeds of the Scottish islands, such as the Shetland, Orkney (North Ronaldsay) and Soay, are this small.[53]

From the same general region, a loose sheep weight account for the

Roberts family has survived from East Sussex.[54] Since the Roberts estates were scattered along the coast from Pevensey to Hastings and inland to Ticehurst and Salehurst, the location of the estate to which this 1687 account related cannot be determined. The sheep were probably killed for the house and were described as 'sheep' and not lambs. Twenty-one legible sheep weights were recorded, with an average weight of 48.95 lb, about 10 lb heavier than the Winchester average. In this series the heaviest weight was 66 lb and the smallest 35 lb, suggesting that only the largest sheep here would exceed 100 lb live weight. The sheep in Sussex may have been of the 'hillish' type, as those described in some Kentish accounts, and were perhaps ancestral to the Southdown breed of the same region in the eighteenth century, or possibly imported western, horned sheep.

Estate records and literary data about fleece weights from the fifteenth and sixteenth centuries remain very rare, but considerably more material has survived for the seventeenth century. The published and freshly examined manuscript material so far available may be summarised in Table 1.

Any average figure for a given location in the sixteenth and seventeenth centuries often concealed a considerable year-on-year fluctuation in yield. Since some of the figures were only derived from a single year's accounts and not from a series, much of the data presented in Table 1 must be regarded as susceptible to a considerable margin of error. For instance, the wether flocks of the L'Estrange family at South Ringstead near Hunstanton in Norfolk had fleece weights varying between 10.5 fleeces per stone in 1693 and 4.3 fleeces per stone in 1695, although this variation seems far wider than any other and may reflect an element of breed change or replacement going on at this location. A more normal fleece weight variation might be that found in 1481–2 in the Townsend flocks of ewes, whose fleece weights in that year varied from 13.75 per stone at Luckham to 10.5 per stone at Sculthorpe, a variation which can hardly have been caused by different climate since these were values for the same year.[55] In the Romney Marsh data for 1697–9, the average fleece weight varied between 3.57 and 4.83 lb.[56]

The fleece weight data for Sussex and Hampshire are broadly consistent with the low carcass weights recorded there and support the view that seventeenth-century southern counties sheep were probably rather small animals. Most of the Norfolk data are consistent with light-fleeced sheep and those from Lincolnshire suggest a divergence between smaller, wold sheep and large, marsh pasture animals. Most of the remaining material comes from the Midland counties and gives a variable picture, clouded by the overstatements of some of the literary sources.

It is difficult to find objective evidence for the increase in size of sheep which has been postulated in the eighteenth century, as there is so little

Breeding sheep

Table 1: *Fleece weight data by county, fifteenth to seventeenth centuries*

County	Date	Fleece weight (lb)	Location
Norfolk	1481–2	1.0–1.9	Fakenham region, N. Norfolk[a]
	1490–5	1.0–1.6	Fakenham region, N. Norfolk[a]
	1490–1510	0.9–1.3	Norwich Cathedral Priory[a]
	1545	1.6	Swaffham[a]
	1548	1.1	Swaffham[a]
	Mid-1550s	1.2–1.4	North-West Norfolk[a]
	1626	1.6	Fakenham region[a]
	1637	1.0–1.4	Fakenham region[b]
	1630–60	0.8–1.3	West Harting[c]
	1658–1726	0.8–1.5	Ewes at Fakenham[c]
	1674	1.8	Wethers at Fakenham[c]
	1685–6	*c.* 1.9	Hockwold[d]
	1693–5	1.1–3.3	Hunstanton N. coast[c]
Lincolnshire	1630s	2.5–3.5	Doncaster market[e]
	1660	4.6	Belton nr. Grantham[f]
	1665	4.8	Belton nr. Grantham[f]
	1665–74	2.5–3.2	S. Ormsby, Lincoln Wolds[g]
	1672	2.0–3.5	555 fleece sale[h]
	1696	7.0–9.0	Grimsby and Wainfleet marshes[i]
Oxfordshire	1586	3.5	Oxfordshire growth[j]
	1598	4.1	Ditchley, Oxfordshire[k]
	1621	4.0–7.0	Growth of Oxfordshire[l]
Buckinghamshire	1598	4.1	Quarrendon[k]
	1607	2.0–3.0	Vulgar growth of Bucks[m]
	1621	4.0–7.0	Growth of Bucks[l]
Northamptonshire	Late 16th century	2.8 3.0–3.5	Isham, between Northampton[n] and Market Harborough
	1607	2.0–3.0	Vulgar wool of Northants[m]
Leicestershire	1607	2.0–3.0	Vulgar wool of Leicester[m]
Worcestershire	1621	4.0–7.0	Worcestershire growth[l]
Warwickshire	1636	3.5	Milcote, S. Warwicks[o]
Yorkshire	1641	2.0–2.8	Elmswell, East Riding[p]
Kent	1607	4.0–5.0	Romney Marsh[m]
	1685	3.2	Romney Marsh[q]
	1697–9	4.2	Romney Marsh[r]
Berkshire	1610–20	2.8	Harwell, Berks Downs[s]
Sussex	Mid-17th century	1.5–3.0	Funtingdon, West Sussex[t]
	1621	2.0–3.0	Sussex growth[l]
Hampshire	1558–60	1.3	Exton, Hampshire Downs[u]
	1621	2.0–3.0	Hampshire growth[l]

[a] K.J. Allison, 'Flock management in the 16th and 17th centuries', *EHR*, 2nd series, 11 (1958–9), 105.

material upon which to construct any chart of the size variation in sheep over time. Evidence from the last thirty years of the eighteenth century can, however, be used to check the accuracy of the journalistic descriptions of the size of animals. On the whole it seems that reporters were reasonably accurate in their assessment of the average regional carcass and fleece size and they did not show any tendency to overstate size, which might have been expected since authors and farmers were at this time still intent on recording the exceptional animals which they had bred or fattened.[57]

Table 2 contains a series of weight data for sheep carcasses derived largely from farm accounts or, if drawn from literary sources, records made of reasonable samples of actual animals. Until far more eighteenth-century farm accounts have been sampled over a greater geographical range, any conclusions drawn on the basis of these figures must be extremely tentative.

Notes to Table 1 (*cont.*)

[b] K.J. Allison, 'The wool supply and the worsted cloth industry in Norfolk in the 16th and 17th centuries, part 1: the sheep husbandry', unpublished Ph.D. thesis, University of Leeds (1955), pp. 196, 221, 227, 228, 245, 262, 285.

[c] Allison, 'Wool supply', pp. 294, 298, 301, 305.

[d] HFRC Reading, Norf 17/1/1.

[e] P.J. Bowden, 'Wool supply and the woollen industry', pp. 44–58.

[f] Lady E. Cust, *Records of the Cust family, series 2: the Brownlows of Belton, 1550–1779* (London, 1909), pp. 80, 82, 87–8, 93–4.

[g] LAO, Massingberd-Mundy Deposit MM 6/5, nos. 2, 5, 8, 17, 18, 19, 27, 28.

[h] Bowden, 'Wool supply', p. 49.

[i] Merret, 'Account of several observables in Lincolnshire', pp. 343–4.

[j] Bowden, 'Wool supply', p. 48.

[k] Bowden, *Wool trade*, p. 32.

[l] F.W. Notestein, F.H. Relf and H. Simpson, *Commons debates 1621*, vol. 7 (New Haven, 1935), p. 499.

[m] C.E. Raven, *English naturalists from Neckham to Ray* (Cambridge University Press, 1947), quoting from E. Topsell, *A history of four-footed beasts (1607)*.

[n] M.E. Finch, 'The wealth of five Northants families, 1540–1640', *Northants Record Society*, 19 (1956), 46.

[o] KAO, U269 Sackville MSS E226/10 (16).

[p] C.B. Robinson (ed.), 'Rural economy in Yorkshire in 1641. Being the farming and account books of Henry Best of Elmswell in the East Riding', *Publications of the Surtees Society*, 33 (1857), 24.

[q] W. Carter, *A summary of certain papers about wooll . . .* (London, 1685).

[r] HFRC Reading, Ken 19/1/1.

[s] G.E. Fussell (ed.), 'Robert Loder's farm accounts, 1610–1620', *Camden Society Publications*, 3rd series, 53 (1936).

[t] Bowden, *Wool trade*, p. 34.

[u] M. Zell, 'Accounts of a sheep and corn farm, 1558–1560', *AHR*, 27 (1979), 122–8.

Table 2: *Sheep carcass weights for the eighteenth century*

Location	Date	Average weight (lb) whole and quarter	Number in sample
Coton Hall Farm, Bridgenorth,	1751	36.8 (9.2)	13
Shropshire	1761	49.3 (12.3)	6
	1763	49.8 (12.4)	12
		35.2 (8.8)	10 Welsh
	1765	34.9 (8.7)	12 Welsh
		38.2 (9.6)	9[a]
Foremark, Derbyshire	1768	66.8 (16.7)	30[b]
Ombersley, nr. Droitwich,	1761	76.4 (19.0)	20
Worcestershire	1764	74.0 (18.5)	57
	1788	56.0 (14.0)	19
	1794	49.9 (12.5)	16
	1797	51.9 (13.0)	25[c]
Worcestershire sheep bought in that county	1796	85.0 (21.3)	19
New Leicester sheep	1796	81.0 (20.3)	19[d]
Lincolnshire long-woolled sheep killed in that county	1791	146.4 (36.6)	12[e]
Holkham Hall, Norfolk Horn sheep	1731–7	55.0 (13.8)	unknown
Castle Howard, Norfolk Horn sheep	1781	56.0 (14.0) 60.0	15[f]
Richard Grenville, Bucks	1741	53.0 (13.3)	8[g]
Chichely, nr. Newport Pagnell,	1782	65.3 (16.3)	53
Bucks	1785	57.0 (14.3)	67[h]
Shardeloes Park, nr. Amersham, Bucks	1800–1	84.1 (21.0)	98[i]
Knole Park, Kent	1713	89.8 (22.5)	18 wethers
		67.7 (17.0)	23 ewes
	1719	69.1 (17.3)	14
	1721	85.0 (21.3)	38 wethers[j]
Waldershare Park, nr. Deal, Kent	1748–68	56.8 (14.2)	503[k]
Improved Southdowns in Essex	1793	93.8 (23.5)	5 wethers
Improved Southdowns owned by the Duke of Bedford	1796	82.0 (20.5)	19
Wiltshire Horns owned by the Duke of Bedford	1796	96.0 (24.0)	19[l]

[a] HFRC Reading, Sal 5/1/1.
[b] Berks RO, D/EBu A5.
[c] Bucks RO, Drake MSS D/D/A 170.
[d] Duke of Bedford, 'Experiment on the comparison of four breeds of sheep', *Annals of Agriculture*, 26 (1796), 412–37.

In the majority of instances samples of animals were killed for consumption by the owners of the flocks concerned and were mature sheep as opposed to lambs. The data from Knole Park show clearly that this is potentially confusing since the balance of ewes and wethers in any sample will influence the final average strongly. At Knole, the animals killed in 1713 were specified as ewes and wethers and entered separately. The wethers averaged nearly 90 lb dead weight but the ewes weighed less than 70 lb, only 75% of the size of the wethers. All the unspecified figures are therefore subject to a wide margin of error in any comparison that may be made between them.

The only weight data from the seventeenth century with which these eighteenth-century values can be directly compared were the two sets derived from Hampshire and East Sussex. The average Winchester dead weight was about 40 lb while that in East Sussex was about 49 lb. If these figures were representative of the sheep in these regions in the late seventeenth century, it would seem that the best of these breeds were much larger sheep by the end of the eighteenth century. The improved Southdown killed out at around 80–90 lb.[58] The western, horned type at Winchester was probably the older and smaller variant of the Wiltshire Horn, the old Hampshire. If this breed is assumed for Winchester, then the best large western Horns in the late eighteenth century were of the order of twice the size. These differences appear to be in sharp contrast to the more static weights of cattle over this hundred-year period. However, much of the weight gain of the Southdown and the Wiltshire by the end of the eighteenth century had probably been achieved only in the last thirty years of the century. Certainly the Wiltshires were recorded as having been enlarged by crossing with bigger rams during this time and the Southdowns recorded here were the elite of the breed.[59]

The rest of the data derives from regions which were not noted for the purity or quality of their flocks. In Worcestershire, the imported Warwick and Gloucester longwools disputed with the indigenous fallow and heath

Notes to Table 2 (*cont.*)
[e] Sir J. Banks, 'Account of twelve Lincoln sheep', *Annals of Agriculture*, 15 (1791), 357–61.
[f] Fussell and Goodman, 'Sheep and wool production'.
[g] HFRC Reading, Buc 11/1/5.
[h] Bucks RO, Chester MSS D/C/4/30.
[i] Bucks RO, Drake MSS D/DR/2/118.
[j] KAO, U269 Sackville MSS A45/1 and A302.
[k] KAO, U471 Waldershare Park MSS A60.
[l] Duke of Bedford, 'Four breeds of sheep', pp. 431–3.

sheep throughout the century and it would appear from the Ombersley material that some farmers in the north-west of the county, having experimented with the longwools earlier in the century, had reverted to a fallow or mixed flock by the 1790s. In the 1760s the Ombersley sheep were clearly of longwool proportions, killing out at about 75 lb, most of them perhaps deriving from Gloucestershire since some of the sheep in this account were specifically referred to as 'Cheltenhams'. They were presumably of the same type as bought by the Duke of Bedford in the county thirty years later, when they seem to have been perhaps 15% larger, at about 85 lb. However, in the 1780s and 1790s the average size of the sheep kept at Ombersley dropped sharply to about 50 lb, a weight compatible with the pasture grazing of small, heath sheep kept in the lowlands.

North of Worcestershire in Shropshire, not far from the Long Mynde and Morfe Common, with their persistent heath and common breeds, the Coton Hall flocks were composed of small sheep. Those specifically designated Welsh were very small indeed at about 35 lb dead weight, while their other sheep seem to have averaged about 50 lb, a size compatible, as at Ombersley, with well-kept heath or heath-cross sheep. At Foremark in South Derbyshire, Sir Robert Burdett's house-killed sheep were of only middle size, probably a mixture of imported grazing sheep, rather than the high-class flock of New Leicesters that this improving livestock breeder was reported as possessing in 1771.[60] However, Burdett's sheep were certainly within the range of weights of longwool pasture sheep at 67 lb since the ewes in Knole Park in 1713 were of the same general size. From the size of the wethers at nearly 90 lb at Knole, these were clearly Romney Marsh longwools. Throughout the eighteenth century, in those regions of Kent where West Country sheep were not imported, it was the habit of the local farmers to graze marsh and lowland-bred longwools, even on the exposed and light-soil hills of East Kent (a parallel with the ability of the Cotswolds and the Lincolnshire Wolds to support sheep of purely pasture type). The wethers at Knole in 1721 were of the same general size, and the largest weights recorded of what were evidently longwools in the Midlands, apart from those of the Lincoln marsh sheep, came into the same size bracket of 80–90 lb. It would appear that the New Leicester was of the same order of size and that by the end of the century the essentially light-soil folding types, the Southdowns and the Wiltshires, were able to achieve the same dimensions. The New Leicester was said to be a small sheep in pasture longwool terms, but this seems to have been only by contrast with the massive marsh Lincolns and Teeswaters. The general run of Midland longwools would appear to have been about 85 lb dead weight, and this size was reasonably consistent throughout the century. The main increases in size were, as expected, among the upland and heath sheep and also among the marsh longwools. At about 145 lb the Lincolns recorded by Sir Joseph Banks in 1791 were clearly huge animals, almost twice the size of typical Midland pasture sheep.[61]

The Norfolk Horns recorded here were of only moderate size and perhaps belonged to the older, smaller type rather than the enlarged variant which Marshall noticed in the 1780s. At Shardeloes and Waldershare Park, the sheep were probably of mixed origin and bought in from elsewhere for grazing. This was certainly the case at Waldershare, where Romney Marsh and Dorset sheep were both grazed in the mid-eighteenth century.

The judgement about the size of sheep can to a certain extent be supplemented by the more abundant data on fleece weight. On the whole, though, this is of little help as it appears that the axiom that larger sheep bore heavier fleeces certainly did not apply in the case of the improved upland and heath breeds. An increase in their size may not have been accompanied by any associated increase in fleece weight, although it may well have coarsened the quality of their short wool. In the case of the pasture longwools it seems to be true that the massive Teeswater and Lincoln breeds did indeed ultimately have very large fleeces, but it seems unlikely that any great enlargement of Midland Plain carcass size or fleece weight occurred during the eighteenth century. Any increase in meat or wool output on the Plain was probably due to the wider spread of the larger pasture sheep, while in the marshes of the east coast it seems that size increase was achieved by selective modification of the breed. On the light soils, for which detailed descriptions and farm accounts have not yet been adequately surveyed, the increase in output was probably due to a combination of both the development of larger animals in some places by accidental or deliberate selection, and the spread of these animals into other areas.

In Table 3 the figures have been drawn where possible from farm accounts or wool yield data quoted in the literature, but no estimates or mere journalistic descriptions have been included. In Kent the wool-weighing system was the 60 lb draft and the 240 lb pack, while in Sussex a 32 lb tod or weight was employed. In Lincolnshire, Berkshire and elsewhere, a standard 28 lb tod has been assumed throughout, although there is some evidence that in North Lincolnshire a 29 lb tod was customary.[62]

The extensive material for Kent and Sussex in Table 3 seems to be reasonably consistent. At Burmarsh in the late seventeenth century (see Table 1) the average weight of Romney fleeces was 4.18 lb and on the same farms in the early eighteenth century the weight was not significantly different at 4.22 lb. The other entries for the Romney area in the table suggest that the fleece weight may have crept up slowly during the eighteenth century from a little above 4 lb to something under 5 lb. The Kent marsh sheep were also found in the coastal marshes of Sussex and the fifth

Breeding sheep

Table 3: *Sheep fleece weights in the eighteenth century*

	Location	Date	Average fleece weight (lb)	Sample size (no. of fleeces)
1	Burmarsh and other Romney Marsh locations, Kent	1700–19	4.22	Very large[a]
2	Unknown location in Kent, but clearly marshland	1767–8, 1770	4.67	329[b]
3	Romney Marsh flockmasters	1773–93	5.17	Very large
4	Rye Custom House, Kent	1792	4.62	60,983[c]
5	Sussex coast, west of Shoreham	1769	4.06	40[d]
6	Little Chart, nr. Ashford, Kent	1727	2.27 (lambs + sheep). 3.00 (sheep only)	522[e]
7	East Sutton Place, Kent	1739	3.10	45[f]
8	Mersham-le-Hatch, nr. Ashford, Kent	1746	2.80 (West Country sheep)	97[g]
9	Waldershare Park, nr. Deal, East Kent	1777–9	3.02	1210[h]
10	Ticehurst, Kent/Sussex border	1787	2.55	269[i]
11	Coombe-in-Hamsey, East Sussex	1769–70 1798–9	2.41 2.83	1434[j] 44
12	Glynde Place, East Sussex	1767	2.23	35 ewes[k]
13	Hastings Custom House, East Sussex	1782–6	2.44	Large
14	Eastbourne Custom House, East Sussex	1768–86	2.25	Large
15	Earl of Ashburnham at Bexhill, East Sussex	1789–93	2.24	Large
16	Newhaven Custom House, East Sussex	1773–92	2.41	Large
17	Shoreham, East Sussex	1793	2.59	578
18	John Ellman's improved Southdowns	1782–8	2.34	Large
19	Southdown sheep at Goodwood, nr. Chichester, West Sussex	1773–82	2.78	Large[l]
20	Bury Farm, Sussex/Hampshire border	1755	2.20	650[m]
21	Arundel Custom House, West Sussex	1772–92	2.88	Large[n]
22	Chilgrove, West Dean, nr. Chichester, West Sussex	1792–1801	2.98	2751[o]
23	East Lockinge, Berkshire	1717–19	3.09	281[p]
24	Unknown location in Berkshire, probably Downs	1771	1.78 (Knott and horned)	39[q]
25	Richard Grenville, Bucks	1742	7.98	32[r]

Table 3 (*cont.*)

Location	Date	Average fleece weight (lb)	Sample size (no. of fleeces)
26 Ombersley, nr. Droitwich, Worcestershire	1761 (estimate)	1.58	110[s]
27 Lawn Farm, somewhere in Hertfordshire	1761	4.62	296[t]
28 Wardley, Rutland	1783–4	9.63	1245
	1803–4	7.65	1319[u]
29 Lincolnshire farmer, probably Wold region	1706	5.71 (+ pack of small wool)	135[v]
30 Ormsby, Lincolnshire Wolds	1718	5.92 (large wool)	175[w]
		3.12 (small wool)	55
31 Outer marsh grazier, Lincolnshire	1743	14.33	688[x]
32 Kirkleatham, Teesvale, nr. Redcar, imported Lincolns	1770	6.60	141[y]
33 Keelby area of North Lincolnshire Wolds and Humber marshes	1775–86	7.70	1152[z]
34 Gedney, Lincolnshire, South Holland marshes	1782–92	8.33	4191[aa]
35 Breeder of Lincoln/New Leicester cross	1796	8.70	1518[bb]
36 Nettleton and Thornton, North Lincolnshire Wolds and marshes	1805–6	7.15	724[cc]

[a] HFRC Reading, Ken 19/1/1.
[b] KAO, U301 Miscellaneous deeds E6.
[c] A. Young, 'A tour through Sussex, 1793', *Annals of Agriculture*, 22 (1794), 171–334, 494–631.
[d] G.W. Tomkins (ed.), 'The Tomkins diary', *Sussex Arch. Collections*, 71 (1930), 22.
[e] KAO, U386 Darrell MSS A2.
[f] Melling, *Kentish sources*, vol. 3, p. 53.
[g] KAO, U951 Knatchbull MSS F18/2.
[h] KAO, U471 Waldershare Park MSS A14.
[i] ESRO, Dunn MSS 38/7.
[j] ESRO, Schiffner archive 3570.
[k] ESRO, Glynde Place archives 2937.
[l] Young, 'Tour through Sussex'; Young, *Agriculture of Sussex*, pp. 47–74 *passim*.
[m] WSRO, Add. MSS 9432.
[n] Young, 'Tour through Sussex'.
[o] WSRO, MP 1478.
[p] HFRC Reading, Ber 43/2/2.
[q] Berks RO, D/EX 62/2.

(*Notes cont. p. 186*)

entry (for the Shoreham coast) was also clipped from sheep which were of the same type. Elsewhere in Kent, at Deal and in the Weald (entries 6 to 10), the general pattern was of grazing mixed breeds of western and marsh sheep, so that the fleece weight averages of $2\frac{1}{2}$ to 3 lb seem to be a reasonable reflection of this situation. Throughout East Sussex the Southdown breed was hardly challenged in the eighteenth century and the extensive records quoted by Young for this region show a consistent $2\frac{1}{2}$ lb fleece clipped. John Ellman's improved sheep (entry 18) do not seem to have had a significantly heavier clip in the 1780s although the Southdown of the early eighteenth century probably had a heavier fleece. Except where it was specifically stated that Southdowns were involved, most of the sheep of West Sussex would have been of West Country type, and the data hint that the fleece of these sheep may have been generally nearer 3 lb than $2\frac{1}{2}$ lb. This yield difference was hardly significant, although there was certainly a difference in the quality and grading of these two fleece types for the clothing trade.

Entry 23 for Berkshire in the early eighteenth century is also consistent with western, horned sheep of the Wiltshire or Hampshire type at a fraction over 3 lb, while entry 24 seems suspiciously light. However, the farmer, Thomas Johnson, kept a mixed flock of horned, presumably western sheep and of polled, Berkshire sheep. The low fleece weight rather suggests that this breed was a very low-yielding down sheep. There is some evidence that Johnson was making an effort to improve the size of his sheep by paying for better rams. Entries 15 and 16 are unclear in this respect. The Ombersley figure of 1.58 lb (entry 26) would be consistent with the fleece of a local heath or common sheep, but the evidence for house-killed animals has suggested (see Table 2) that at this time the estate was stocked with pasture longwools. The only rational explanation is that there were different flocks of sheep involved and that the fleece figures came from a separate flock of

Notes to Table 3 (*cont.*)
r   HFRC Reading, Buc 11/1/5.
s   Berks RO, D/ED E48C.
t   HFRC Reading, Hert 1/1/1.
u   R. Parkinson, *General view of the agriculture of the county of Rutland* (London, 1808), pp. 129–30.
v   J. Smith, *Chronicum rusticum commerciale or memoirs of wool . . .* (London, 1747), vol. 2, pp. 463–4.
w   LAO, Massingberd-Mundy Deposit MM 5/2.
x   Perkins, *Sheep farming in Lincolnshire*, p. 8.
y   Young, *Northern tour*, vol. 2, p. 136.
z   LAO, Dixon MSS 4/1.
aa LAO, Miscellaneous deposit 150/1.
bb Young, *Agriculture of Lincolnshire*, p. 351.
cc LAO, Dixon MSS 5/2/2.

heath sheep. A similar anomaly occurs in Richard Grenville's Buckinghamshire accounts (entry 25), since the carcass data suggest the sheep were only of moderate size, while the fleece data for the following year, if no mistake has been made, suggest a pasture sheep of the largest size, since an 8 lb fleece would certainly be the heaviest that a mid-eighteenth-century pasture longwool of the Midland Plain could sustain. At Wardley in Rutland, in the 1780s (entry 28), average fleece weights of $9\frac{1}{2}$ lb were obtained, but probably only following the spread of sheep with very long wool developed in Lincolnshire in the latter part of the century. In the early nineteenth century these had dropped back to about $7\frac{1}{2}$ lb through the crossing of the old Rutland stock with the smaller and lower fleece-yielding New Leicester. The Hertfordshire figure for 1761 of 4.62 lb (entry 27) probably represents fleeces from a mixed flock of grazing sheep, either separate types introduced as grazier's stock, or a mixed breed produced from crossing long-woolled pasture sheep with one of the arable, folding types found in or imported into the region.

Most of the data from Lincolnshire fit into the expected pattern as well. The records early in the century drawn from the lower slopes of the Wolds show sheep of two types, as was characteristic of the Wolds in the seventeenth century: the larger-fleeced sheep with 5 to 6 lb fleeces and the type from the higher Wolds with a smaller fleece of about 3 lb. There is no primary evidence to support it, but there is a reasonable hint that the salt marsh region of the coast to the east of the Wolds was already producing sheep with fleeces longer than those in the late seventeenth century, and these marshes continued to be where the very large and heavy-fleeced Lincolns grazed. The weight difference was probably compatible with the better marsh grazing of the essentially middle-woolled type of the lower Wolds. Early in the century the marsh graziers did not breed their own sheep but bought them from the specialist breeders of the lower Wolds, who were unable to finish them to such a great weight. As the farms of the Wolds were enclosed and farmed more productively later in the century, the Wold farmers were able to grow their sheep on to finished weight and there was not so much excess for the marshes. As a consequence, the fleece yields of the lower-Wold farms increased and the marsh graziers themselves began to take an interest in the breeding of sheep. In terms of the breeding of stock selected for specific purposes this change meant that the finishers and breeders of the stock became the same people, so that it was possible to adopt a more rational breeding policy than had been the case when the breeder had no concern with the final quality of the animal that he was producing. The marsh graziers, who had always considered the wool clip their most profitable product, were able, from the mid-eighteenth century onwards, to select for animals with really massive fleeces, at the same time producing the characteristic giant, watery Lincoln carcass. The stimulus to step up the fleece yield was provided by the continually poor return for the growth of longwool from about 1730 onwards. The Lincolnshire graziers made all kinds of attempts to counteract or mitigate the effects of this relative fall in the real price of their product, one of which was to increase the amount of the product that any one sheep would grow.[63]

From the 1770s onwards, therefore, reports began to appear of truly massive fleece weights on outer marsh sheep in North Lincolnshire. At the same time, and for the same general reasons, the yields of lower-Wold and mixed-Wold and marsh sheep began to increase. These lower-Wold sheep were carrying 8 or 9 lb fleeces at the end of the century, compared with 5 or 6 lb ones at the beginning. By very late in the century, reports of 14 lb fleeces on outer marsh wethers were not uncommon and there may have been some whole flocks with average fleece weights of this size, although the literary evidence does not suggest this could have been all that common.[64] In this context, entry 31 in Table 3 looks suspicious since it suggests that the sheep of the outer marsh were able to achieve average fleece weights of 14 lb some fifty years before this could be expected. The value is drawn from a modern source, quoting a newspaper report of the 1840s about a Tod Bill of a hundred years earlier.[65] Under the circumstances it would seem more reasonable to assume that a transcription or calculation error has crept in somewhere here, rather than that such massive fleeces existed at the beginning of the period when the outer marsh graziers began to breed their own sheep.

Sheep breeding policy in the seventeenth century and before

Evidence for any improvement in the productivity of sheep farming in the seventeenth century is scanty. The trend towards pasture rather than field sheep in the Midland counties and the large sheep kept on the coastal marshes were the only obvious signs of such positive change. The extent to which the sheep breeders were responsible for any improvements rather than the changes being the consequence of the mere spread of more productive general systems of husbandry is unclear. Some discussion of sheep management and breeding policy is called for here to assess its potential for livestock improvement.

The agricultural literature of the period was usually uninformative on the subject. Fitzherbert made no mention of selective breeding at all and gave no advice on choosing rams or ewes for breedstock. It is clear that when conditions were regarded as ideal for sheep husbandry, the sheep were allowed to operate as an uncontrolled flock and hardly interfered with at all. Where conditions were less than ideal, management of the flock was necessary to ensure proper lamb growth and the receptivity of the ewe at tupping time. The inference must be drawn that in those flocks which were best provided for environmentally, the sheep were allowed to operate under natural conditions with no significant artificial selection at all. Combined with Tusser's advice to geld all ram lambs as soon as they fell, presumably leaving only a random group of entires for breeding, it is evident that on the better-nourished pastoral flocks no selective breeding was practised in the sixteenth century.[66]

These sixteenth-century sources gave little evidence of the markets for sheep, but Tusser evidently regarded it as worthwhile to fat old sheep from outside the home flock in August on any spare stubbles which might be available.[67] Mouffet reported that the best mutton came from carcasses not above three or four years old, which suggests that even if the primary market was for culls and old mutton, meat that was just about fully mature (three years old in a seventeenth-century context) was available and should be chosen by the consumer. He also noted that hill-fed sheep had by far the best flavour, while those from the Somerset and Lincolnshire marshes were already considered to be fatty, poorly flavoured and unwholesome, as they were to be increasingly regarded by discriminating palates in the eighteenth century. The slender evidence suggests that in the sixteenth century fat lamb was probably not available.[68]

Markham described the breeding and management of sheep in the early seventeenth century. He believed the major regional characters of sheep to be a product of the environmental conditions rather than derived from any innate quality. He certainly believed that soil colour determined the colour of wool and considered red soils to be preferable in this respect. In general, large (here referred to as 'biggest boned'), very well fleeced sheep should be chosen, with greasy, close, soft and well-curled wool, which should cover the belly as well as the rest of the body. He believed that the desirable fleece type would correlate well with a good butcher's animal, in the sense of fatting speedily when culled out for that purpose, presumably at maturity. In breeding, the ram was to be considered before the ewes with respect to shape and general properties because he would influence the form of all the offspring, whereas the ewes only contributed to one each. His general advice on the best points for such a ram reads like a summary of several classical authors, with an emphasis on the points of the head. The ewes should be selected on the same principles, should be from the same soil and locality and should have the same shape and properties. A pure-breeding policy of mating animals all of the same type was clearly envisaged, and any mixing or crossing of strains was to be avoided.

At Michaelmas male and female lambs were to be separated and the majority of the males gelded. Those preserved for rams should be the 'worthiest', implying that the most vigorous would make the best tups and would therefore beget good lambs in the next generation.[69] Therefore, although Markham said he believed an animal gained most of its desirable properties from its environment, the rams to be used for breeding should nevertheless be carefully selected. The advice to select the ram at six months would allow some judgement of its phenotype to be made and represented an advance on the earlier advice of Tusser and Fitzherbert.

But the most famous account of sheep husbandry in the seventeenth century was that composed privately by Henry Best of Elmswell in the East Riding of Yorkshire, in 1641.[70] Best's description of sheep management has been taken by most authors to represent the standard methodology of the sixteenth and seventeenth centuries and has been used as a basis for many of the accounts of sheep husbandry found in the literature.[71] Best inherited his farm in 1617, moving from the Braintree region of Essex and life as a shopkeeper in London. Ryder suggested he probably had a considerable incentive as a non-farming outsider, to compose a work of this nature describing phenomena which were unfamiliar to him,[72] perhaps a parallel to the similar detailed account of local husbandry written at the end of the century by Edward Lisle in Hampshire.

Best's book concentrated on sheep husbandry and the cultivation of corn and hay and little reference was made to cattle. He appeared to operate his farm as a sheep–corn enterprise. The inventory of John Best, Henry's nephew, at Elmswell in 1668, showed that there were then cattle on the farm worth £58 5s, horses worth £48 10s and a flock of 250 sheep, valued at £75 6s. Cattle did not seem such a minor component of the system.[73] What was missing from this inventory was the value of the corn. Harwood Long has shown that the Yorkshire Wolds were sheep–corn country in the seventeenth century, most of the rest of the county being dominated by cattle. Even on the Wolds the number of cattle kept was not significantly different from the numbers maintained elsewhere, but the value of sheep, corn and draught horses together greatly exceeded their value and showed the strong arable bias of the Wolds.[74] Best therefore seems typical of the general practice of his region.

Best's *Accounts*, published with his *Farming Boke*, show a sheep flock stabilised at around 150 to 200 ewes from the mid-1620s onwards. The lamb numbers recorded in June suggest the balance of ewes and wethers in the flock was about equal and the expected lambing percentage was of the order of 80–90%. On the practical events of breeding, Best recommended running the tups with the ewes all year; allowing them to mate at will in the fields at the breeding season. Provided there was sufficient winter feed from the summer hay crop and there were some sheltered closes for lambing down on the farm, he believed that biologically regulated tupping at

Michaelmas was best, giving a lambing season in February and early March. Many did not allow this practice locally but held their rams away from the ewes until St Luke's Day (18 October) so that the earliest lambs would not fall until mid-March, to allow the lambs a better birth season and to be nearer the spring grass flush. One justification for Best's system was that earlier production of summer fat lamb gave better prices; thus a fat-lamb market existed here, where one might have expected to find evidence only of a leanstock trade for fatting elsewhere.[75] Best claimed a good tup should be capable of servicing between 30 and 50 ewes and the standard number of rams in the flock seems to have been four for every long hundred (six score, or 120).

Matings seem to have been completely uncontrolled, much as Fitzherbert had recommended. The rams were merely run with the whole ewe flock and allowed to fight and build harems of their own during the breeding season. The only precaution seems to have been the separation of riggons and young tups into well-fenced enclosures to keep them out of harm's way and prevent them from mating. The rams themselves were 'most unruly and raininge' and 'will fight cruelly one with another'. Best gave no information on how the tups were selected from the male lamb population, although considerable advice was forthcoming on the timing and mechanics of castration. This was done when the animals were about six weeks old, so the judgement of which animals to keep for rams must have been more or less random. However, Best, following Markham, was well aware of the influence of a poor tup on a flock since it fathered so many lambs. He said that judicious sheepmen went to considerable lengths to obtain good tups but did not elaborate on how they did this.

His specification for a good tup was brief. He should be large and 'well-quartered' (which presumably meant having a good conformation for the butcher), with good, smooth wool, and should be long-tailed and polled. In age he should be over two-shear (more than 27 months) but not above five-shear, and he gave his reasons for this age-range in terms of natural philosophy. Young rams were too hot, so that the scab resulted, while older ones were too dry and 'their radicall moisture is wasted'. He was also most insistent that the ram should have both his stones descended and not be a riggon, where only one had descended, or a close tup, where both remained in the body cavity. His insistence upon this and his emphasis on how to distinguish close tups from wethers suggests that these faults were reasonably common. He also clearly believed that riggon or close tuppedness was hereditary, for it was in the context of guarding against the eventuality of allowing a riggon to breed that he used what he called 'an experienced adage' that *omne animal generat sibi simile*, stressing that he put more faith in the breed and ancestry of a flock of sheep than in the influence that the environment could have upon them. Unlike Markham, and despite his mention of the fat-lamb trade, he did not specify clearly that good or fast fatting properties were desirable or should be selected for.

Good sheepmen were as careful about their ewes as they were about their tups. Selection was once again based on their general phenotype. A well-quartered, well-stapled ewe with a tall, straight lamb was worth twice as much as a short, runtish ewe with a short-legged lamb at foot. This preference for the former was justified in that three times the fleece yield would be obtained off the better ewe and its single lamb would sell better for less cost than two of the inferior ones, implying a faster growth rate and 'kindlier' fatting, while the ewe herself would be worth up to 1s 6d more in the market. It seems sheep farmers were prepared to pay considerably more to hold good stock, believing their money well spent, a first prerequisite to any pattern of general improvement, although there is little evidence of controlled matings as a source of such improvement. Best gave more 'points' in ewe selection than he did for the ram, showing that Yorkshire sheep farmers were aware of the maternal influence on the offspring and its importance.

Like Markham, Best emphasised the conformation and profit to be expected from the butchered beast as well as the traditional profit from the fleece. Both fat-lamb and three-year-old fat-wether production were emphasised. It seemed from his account that the sheep of the Wolds were both horned and polled at this stage, but that polledness was being strongly selected for. Black sheep were valuable at a rate of some four per flock but not more, since the fleeces were invariably hairy. Neither were blue-faced or black-legged tups encouraged. Beyond these general considerations, no other specific details of what breed of sheep he was dealing with emerged, apart from the nature and weight of the fleece. Yorkshire sheep were described by Markham as hairy and Yorkshire fleeces throughout the Middle Ages obtained an ever worsening reputation. Best reckoned to get some 5–7 fleeces per 14 lb stone, equivalent to a washed fleece weight of from 2 to 2.8 lb, and his usual markets for this clip were the West Riding coarse-cloth towns of Leeds, Halifax and Wakefield, which suggests sheep with average quality fleeces rather than poor, hairy ones. Best's field sheep produced a fleece which sold at 8s–9s per stone on average. Those who had pasture wool could demand and be given 25% more, at 10s–11s per stone. These pasture-fleeced sheep might either have been the same as Best's, kept on a pasture regime, or a separate breed of longer-fleeced grass sheep. In either case it is significant that the pasture fleeces were worth more than the field fleeces. The apparently heavier-fleeced sheep did not here give a coarser or cheaper fleece than the equivalent local arable sheep.[76]

In 1651 Samuel Hartlib was most uncomplimentary on the subject of

English sheep husbandry and compared it unfavourably with Continental, and especially Dutch, practice. He accused English flock masters of not being 'curious in procuring the best sorts of sheepe, for greatnesse, soundnesse and fine wooll'. He advocated the import of Spanish sheep to 'mend' our fleece, but seemed to have believed strongly in the influence of environment on the expression of this character since he added that it would probably only do 'for a time' and would deteriorate under English conditions. He reported that Dutch sheep were very prolific and sometimes brought two or three lambs at a time and were large and long-tailed. Despite his criticism of English methods he was forced to admit that foreigners spoke highly of the fineness of English wool.[77]

By contrast, in 1674, John Beale commented that vigorous attempts were being made to improve the race of sheep in England for hardiness, size and fleece weight. Beale gave the strongest literary hint that it was downland and hill farmers who were in the front line here:

[they] do offer very high rates for the largest sheep, Rams especially, that they can procure, and sometimes they buy from foreign parts very large and lusty rams; and find the benefit by the largeness of the descending race.

The upland farmers of small sheep – in many cases with very fine wool – found that the premium paid for the fine fleece was too small to compensate for its very small size, while the superior flavour of upland mutton did not make up for the diminutive size of the carcass. These farmers were seeking out large rams, either variants in the local population or rams from 'foreign' regions, in an attempt to upgrade the size of their stock. Beale said nothing of the actions of the farmers once they had selected these rams since a formal grading-up process by hybridisation would require the breeding of offspring generations to the imported parent type for a period of time.

Beale felt constrained to join the gathering controversy over the preservation of the quality of English wool, observing that the main thrust of improvement was among the raisers of finewools on the uplands, and he also advocated the import of Spanish sheep. Significantly, he used as a model for all the hybridisation experiments he discussed or proposed, the import of Barbary and Spanish stallions and the improvement of English bloodstock. It would be interesting to know whether the farmers themselves had the same model in their minds as a basis for their activities.[78]

If the upland farmers were to import rams from foreign parts, these must have been larger, lowland animals. Some confirmation of this was provided by Fuller, who mentioned that in 1662 good Buckinghamshire breeding rams were fetching the elevated price of £10, well above the going rate for the normal run of sheep or rams.[79]

The contemporary view in the sixteenth and seventeenth centuries was that fleece quality and the overall size of a particular type of sheep were environmentally determined and that the nature of any breed was governed by the conditions in its geographical location. The logical extension of this was to suppose that if a breed was introduced into a new environment it would soon come to resemble the sheep of the new region, as the newcomers were now subjected to the same environmental conditions as the native breed. The most extreme case of this argument was put in an anti-enclosure pamphlet by Clement Armstrong, who stated that Welsh sheep could be changed into staple sheep by bringing them into an English lowland arable environment. But most practising breeders and graziers were well aware that environmental plasticity had its limits and that the innate heredity of the race or strain was more important in the control of animal form; *omne animal generat sibi simile*, as Best said. Concern with sheep quality and heredity may be traced well back into the sixteenth century.

The Spencers of Althorp, in Northamptonshire, were large sheep raisers throughout the whole of that century. The original grazier, Sir John Spencer, stated in 1519 that: 'his lyving ys and hathe been by the brede of cattell in his pastures, for he yr neyther byer nor seller in comon markettes as other grasyers beyn, but lyveth by his owne brede.'[80] He could have followed the precepts of Clement Armstrong, if he had believed them, and stocked his ground with any old sheep whose staple would have improved by virtue of the excellent nature of his pasture. On the contrary, the Spencer stock was carefully bred and had a general reputation for high quality, sufficient to enable the owners to sell choice animals at high prices to others for stocking their farms. If these buyers had believed any version of Armstrong's views they would hardly have wasted time and effort in securing so expensive a breed: they clearly thought this quality was heritable and the stock would maintain its virtue on the new land. For example, in 1602 Lord Burghley wanted 60 ewes and 4 tups 'oute of yor choysest yu can well spare' so that he might stock certain grounds with 'suche as are of good brede'.[81]

More circumstantial evidence for a belief in the inheritance of quality in sheep comes from the action of governments on the question of the export of live sheep overseas. The original prohibition of live-sheep exports appeared in 1424 in Anno 3 Henry VI and was apparently designed to protect the export trade in English wool by preventing the Flemish clothing towns from growing their own wool of English origin.[82] The only permissible live-sheep exports were supposed to be for the victualling of the port of Calais. But since this was the main Continental staple town for English wools, one can see that the prohibition was probably ineffective.

A similar Act was passed in 1565 in Anno 8 Elizabeth.[83] Here, no general reasons were given for the prohibition, but it was probably passed to protect wool supplies for the English cloth trade, then in the doldrums following the crash of 1551. This Act was difficult to enforce, or possibly ignored, for in 1614 the Privy Council had to remind the officers of the English ports that such legislation existed and that they should enforce it.[84] No reasons for the provisions of the Act were given then either, but it certainly represented an element in the rising paranoia of governments and manufacturers about ensuring the supply of cheap English wool for English manufacture. If the Tudor and Stuart legislators had really believed that environment alone controlled fleece quality, why were they so anxious to prevent the export of English sheep?

There are some positive hints that graziers in certain regions were beginning to devise positive programmes of stock improvement through selective breeding. Consider, for instance, Nicholas Toke's Kent stock lists for the period 1616 to 1630.[85] Toke's estates were at Godinton, Milstead and Bonnington in the Kentish Weald, but during the period 1616 to 1621 he owned a piece of marsh grazing ground at Cheynecourt in Romney Marsh and there bred sheep described as 'Marsh', 'Cheynecourt' or, more significantly, 'Cheynecourt breed'. Other entries merely recorded ewes, wethers or whatever, but sometimes these were referred to as 'hillish' sheep. In his stock lists Cheynecourt ewes were usually valued at 12s while other, unspecified ewes were only worth 10s. But in the early years at Cheynecourt 'rams' and 'rams for store' were valued at 25s each. Towards the end of the period when he was using Cheynecourt, the average value of the rams fell and from 1624 when he no longer had any marsh ground, the value of the rams dropped to between 10s and 12s. The clear hint is given that incipient Romney Marsh sheep were larger and more expensive than the normal Kent sheep and that some concern with their breeding was elevating the value of breeding rams (see Table 4).

In most years the value of the rams was greatly in excess of ewes and also in excess of that of wether sheep. However, when the large Romney Marsh grazier, William Deedes, was breeding some of his own marsh sheep in the late seventeenth and early eighteenth centuries, although rams were usually of greater value than ewes, the ratio was 1.5 or less and the ram values were about the same as those of wethers, suggesting the reason for any excess value of rams over ewes here was entirely due to the greater size and yield of the male animals of this type at the time.[86]

Another hint, no more, of the possible actions of improving breeders in the seventeenth century comes from correspondence between Sir Thomas Tresham of Rushton Hall, near Kettering in Northamptonshire, and his steward Mr Hilton. The location is close to Althorp. In April 1603 Hilton had gone into Leicestershire to Whetstone, just south of Leicester, to try and sell some sheep. Twenty of the poorest sort were tried in the market at 35s apiece, well in excess of the maximum prices for Romney Marsh sheep at

Table 4: *Sheep valuations of Nicholas Toke*

Location	Date (October)	No. of ewes	No. of rams	Value of rams	Value of ewes	Ram/ewe value ratio
Cheynecourt	1616	264	22	25s	12s	2.10
Cheynecourt	1617	483	22	25s	11s 6d	2.20
Goddenton	1617	9	24	12s	10s	1.20
Cheynecourt	1618	462	32	25s	12s	2.10
		51 (hillish)			10s	
Goddenton	1619	443	49	20s	11s 6d	1.74
Cheynecourt	1620	462	66	18s	12s	1.50
Goddenton	1624	72	3	12s	9s	1.33
Goddenton	1625	87 ewes and rams		10s	10s	1.00
Goddenton	1626	96	38	10s	11s	0.91
		(including tegs)				
Goddenton	1627	132	9	12s	10s	1.20
Goddenton	1629	213(+ tegs)	13	12s	10s	1.20
Goddenton	1630	260(+ tegs)	16	12s	10s	1.20

approximately the same period (18s to 20s for fat wethers). However, the asking price nearly caused a riot and no offers were made above 28s. Hilton only managed to dispose of two to a 'poor foreigner'. One explanation of these asking prices is obviously that Tresham believed he had some very superior stock to dispose of, although if this was the case, the farmers of Whetstone clearly did not agree with him.[87]

Sheep breeding in the eighteenth century: Robert Bakewell, his colleagues and rivals

In the 1740s Robert Bakewell began his successful experimental sheep-breeding programme, which culminated in the production of the New Leicester or Dishley breed. The diffusion of this breed into various parts of the country began on a significant scale in the early 1770s. It either displaced the previous lowland, long-woolled, pasture sheep by complete breed replacement or imported Dishley sires were used to grade up local ewes. The New Leicester was also used to increase the size of arable and hill sheep, usually by the creation of half-breds. The most advanced ideas and achievements in sheep breeding must be sought among the immediate predecessors and contemporaries of Bakewell in the East Midlands.

It would be fair to say that sheep, before the mid-eighteenth century, were a neglected species in terms of 'scientific' attention and improvement. One serious impediment, to which the only exception were the lowland and marsh grazing sheep, was that most flocks on the light soils, commons and sheep walks led a life far less modified and regulated by domestication than the cattle and horses used for dairy, plough, cart and saddle work. Sheep living under semi wild conditions can only survive and prosper if they are familiar with their range, fit into the wild population social structure and have a genetic complement adapted for survival under natural, and not managed conditions.[88] The failure of half- and three-quarter-bred stock on the mountains in the eighteenth century, derived from lowland crosses to increase their size, resulted from the inadequacy of the genetic and physiological adaptations of the lowland breeds in the conditions on the high hills. In the nineteenth century, the impossibility of 'improving' mountain breeds in this way became obvious and the industry evolved a carefully stratified and very elegant cross-breeding system from rigorously maintained pure-bred varieties, adapted to specific sets of conditions. The great range of semi-wild environments upon which sheep are grazed in the British Isles has ensured that this range of types has persisted. The final combination of desired properties cannot be permanently united in one breed but must be re-created at each generation.[89] The eighteenth-century experimental improvers only learnt this lesson through a series of unsuitable crosses and total failures.

The wild condition in which many sheep were kept may also have inhibited the attentions of the profit hungry landowner.[90] There can be little doubt that the gentry were very much involved with the development of fashionable breeds of eighteenth-century horse, dog and fancy poultry. Even where they were not themselves breeders, their interests and the markets which they provided ensured that farmers and breeders who did supply them had incentive enough to increase their own profits with better stock. With one or two notable exceptions, such as Lord Sheffield, the gentry and aristocracy showed little practical concern with sheep husbandry. For instance, Sir John Sinclair and the well-connected supporters of the Society for the Improvement of British Wool did not involve themselves to any great extent in the practicalities of breeding. Exposed hills and cold, early-morning spring lambing pens could have had only limited appeal to civilised improvers, whose whole philosophy was based upon the exertion of control over the environment and the physical comfort that such control implied.

The very large number of sheep types and their evident close adaptation to their various localities reinforced the belief that sheep breeds were merely environmentally induced variants, a prejudice supported by the views of some naturalists that domestic breeds were merely accidental varieties whose type was not permanent or hereditary. Agrarian writers frequently asserted that heredity played no role in the form of domestic animals and that environment was the cause of all variation.[91] Even after the changes wrought by Bakewell and Ellman had been widely diffused, the opinion that the sheep of any region were largely determined by the local conditions persisted. Certainly, in the 1770s, when Anderson wrote a careful analysis of the types of improvement which might be carried out on livestock, it was true that the issue of the pre-eminence of heredity over environment in the control of breed form could not be decided, as no controlled trials had taken place in the case of farm breeds.[92] However, breeds of dog and the Thoroughbred horse already provided models of the permanent inheritance of variant domestic forms. The earliest sheep improvers worked in the lowlands and were less aware than most of the peculiar characters of sheep breeds in small, local regions and were therefore, perhaps, less inclined to believe the environmental argument. It is also true that the animals they were dealing with were remarkably free from distinct fancy features, which so many breeders of local forms regarded as critical to the virtue of their animals.

There can be no question about the pivotal position of Robert Bakewell among these lowland sheep breeders. His influence was far more important here with the dissemination of the Dishley breed than it was in the case of the Midland Longhorn, or his less well publicised activities in pig and horse breeding. Not only was the sheep the earliest species upon which he worked, but it seems clear that none of his contemporaries could claim equal credit for the formation of the New Leicester. They all followed his lead. Even at the end of his career when the Dishley Tup Society was formed in the 1780s, his associates never really managed to challenge his dominance. By contrast, the Longhorn arrived in the East Midlands from its home in the North-West along many routes and into many hands. Bakewell himself did not breed a really successful Longhorn animal until the late 1760s and did so using traditional Lancastrian methods. Conversely, he seemed to have begun his sheep-breeding activities in the mid-1740s and to have developed his ideas during a long period of concentration on this species. By 1770 most of his views on breeding had probably crystallised and his work with cattle after 1770 and with pigs and horses in the later years probably consisted of their application to as many species as possible.

The last fifteen years of Bakewell's life have been adequately described by a number of writers, who based their accounts on the extant records of the man and his closest Midland and Northumberland associates.[93] But the exact nature of his achievement still remains open to doubt. The objectives and methods that he advocated in these later years were well recorded, although whether the methods he actually used were precisely those he expounded, or some subtle and possibly devious variants, needs more consideration. Much more open to doubt are the materials from which he forged the Dishley Leicester, and whether the aims and methods so well publicised later in his career were those he employed in its foundation, or did these theoretical and technical procedures evolve slowly, along with the breed? His later objectives and methods may be briefly summarised.[94]

Plate 20. A shorn New Leicester ram. Lithograph from David Low (1842). The animal shown here had the 'barrel' form which Leicesters were supposed to have derived from Bakewell. This animal was blocky, straight-backed and shown newly shorn to accentuate its mutton conformation. The same straight-backed, barrel forms were depicted in earlier illustrations, such as those shown in Plates 22 and 23. In the Low lithograph, however, the heavy breast region was less emphasised. The original Bakewell Leicester was said to have been very heavy on the forequarter.

(a) *His primary economic objective was the creation of a profitable meat animal* in the case of both sheep and cattle. The sheep fleece and the milk yield of the cow were of comparatively little interest to him, although as a commercial publicist he insisted on trying to show that his sheep were hardy enough for the fold and that his cattle would plough and milk.

(b) *His detailed objectives within the context of meat production were several*:

(i)   He aimed to produce a fast growth rate so that animals could be got ready for market quickly. A higher stock turnover meant a more efficient use of grazing resources.

(ii) He aimed to produce an efficient conversion of fodder into meat. The ideal was an animal which wasted the least of its input in walking, respiration and other wasteful activity and turned the maximum percentage into flesh.

(iii) He aimed to maximise the proportion of saleable meat on the carcass in relation to those products which brought the grazier little profit. Thus he emphasised the value of muscle tissue and covering fat (both edible) at the expense of offal, hide, head, hooves, skeletal bone and internal fat (tallow) which could not be eaten and for which the grazier could not claim a just price. The butcher, however, did sell the tallow, hides and bones. These represented sources of profit for which he gave the grazier little financial return. He was not necessarily happy with an animal which yielded more meat, but for which he was also expected to pay a higher price.

(iv) Bakewell sought to distribute the flesh over the carcass in the most profitable manner so far as the grazier was concerned. In what looked like a mere expedient derived from the given breed characters that his animals possessed, Bakewell emphasised the expensive cuts of the cow on the rear-quarters, but the cheap cuts of the sheep on the fore-quarters, arguing that the wealthy purchased beef and wanted good quality, while the poor bought mutton and required economy. Bakewell's concept of 'flesh' was also debatable. Especially in the animals which he bred towards the end of his life, his main objective was to put on fat, not muscle. Even his admirers at this time had to admit that much of Bakewell's stock was far too fat. The emphasis on fat certainly arose because it was easier to obtain than muscle, especially as Bakewell kept his animals in very high condition and because the whole historical emphasis of Midland grazing skills had been concerned with the choice of a carcass which would put on fat quickly and easily for market. Bakewell's activities tended to exaggerate this tendency to a ludicrous degree. The excessive fatness also assisted in giving an impression of the relative smallness of bone and offal.

(c) *Bakewell's most important technique was the standard and historical one of selecting strongly* for the animals whose appearance or performance he wished to perpetuate, breeding only from those conforming to his view of a good animal. He was undoubtedly prepared to select rather more rigorously than some of his predecessors and contemporaries. As with all domestic breedstock, where one male contributes to many more offspring than each female, his policy was to select more strongly among rams than ewes. Besides the practical logic of numbers, the procedure had the advantage of emphasising the contribution made by males to their offspring, which was probably more favoured as a philosophical position than the alternative of equal contribution or a dominant role for the female, although the anthropomorphic association between mother and offspring going back into classical times may have counteracted this. It is true, for instance, that Bakewell was well aware of the contribution that Webster cows had made to his race of cattle.

Plate 21. The Southdown breed. Lithograph from David Low (1842). The animal shown here was a four-shear ram bred by John Ellman, the founder of the improved Southdown breed. This rapidly displaced the Old Wiltshires and Norfolks in the early nineteeenth century.

Where Bakewell seems to have been far more sophisticated than other agricultural breeders was not in the principle of strong selection, but in the nature of the characters that he selected for. He claimed that he had aimed at economic performance, rapid growth rate and efficient food conversion (physiology rather than mere form), but since he did not have the resources or the equipment for the proper measurement of such performance until late in his career, the strength with which he could have selected for it in the early days must be called into question. There can be no doubt that phenotypic conformation must have formed the basis of his selection programme. He chose animals which looked right, which in his terms meant those that were easy to fat, were thin-legged (since that was the place to observe the bone size in the live sheep) and conformed to the shape which he believed would reflect the best carcass form for flesh and fat distribution. Most of his contemporaries also selected strongly for phenotype, and in the case of those in Lincolnshire, most successfully for the character of wool yield, and perhaps more successfully than Bakewell for his prime character of body conformation. Other breeders also selected strongly for phenotypic characters, as they had done since classical times, but often it seemed to be

only for the obvious and economically insignificant fancy features of colour
and horn form. Since such selection was never consistent, because different
individuals preferred different expressions and the fashion for such points
was constantly changing, the sheep population as a whole remained
variable for these features and allowed for the change in fashion and
preference to go on endlessly cycling round. Such selection often proved to
be very effective, since the general character of fancy breed points could
radically change with fashion in ten or twenty years.[95] Farmers almost
certainly believed such fancy points were correlated with economic
performance and the persistence of fancy points, pure and simple, in modern
breeds show how tenacious this breeding habit is.

(d) *Many breeders almost certainly only half believed that breed characters
were heritable* and may have bred their stock randomly, choosing to fill their
flocks with sheep of a desired type which just happened to crop up among
their home-bred animals and obtaining the balance of such types in the
market place. However, most breeders must have selected their breedstock
in the hope (if not the firm belief) that the offspring would resemble the
parents and that like would beget like. In the case of Bakewell and many of
his contemporaries in the breeding business, the belief was certainly a
conviction that not only would specific and generic characters be hereditary, but mere breed or varietal characters would be as well. Given the mixed
and heterogeneous state of most livestock at this time, this conviction
probably required considerable faith, since in so many cases animals would
not transmit their characters to their offspring with any consistency.
Provided that some guarantee could be had that the animal concerned came
from a race of like animals, rather than being only one form among many
variants, the probability of the conviction being true was enhanced. This
was one of the bases of the pedigree system of recording animal ancestry,
essential in any properly conducted flock or herd where consistent
directional selection was to be exerted. Bakewell certainly paid considerable
attention to pedigree and began to convince himself of some of the more
mystical interpretations that always get put on pedigree data. He believed
that a good animal from a good blood line would stamp its influence on more
distant offspring even if its immediate progeny were undistinguished. To this
extent, certainly later in his career, his selection system was based entirely
on blood line or ancestry in the first instance, and poorish stock from the
'lead' lines was preferred to better animals whose ancestry was less
distinguished, an attitude that was already common among racehorse
breeders. John Ellman also held the same view and would only choose an
animal with a good phenotype it if had been well bred for several
generations.[96]

(e) *The conviction that like begets like would also lead to the inevitable tendency to breed closely from within one's own herd or flock*, once a selected group of desirable animals had been drawn from the surrounding populations. Once the selection of good stock had been made, especially if it was individualistic and based on painstaking observation and analysis (even if much of this was subjective and imprecise), this led to the need to carry on breeding from one's own stock, to remain 'within the same strain'. For reasons which may have been practical, prejudiced or religious, most breeders evidently thought such continuation in a strain would lead to degeneration and so they would cross their flocks with those of their neighbours every few generations, although it is probable that the more enlightened would use only stock from reputable flockmasters and choose only animals which seemed to them desirable and sound, rather than merely crossing at random for the sake of the cross. Bakewell, like some horse breeders and writers before him, did not believe that inbreeding within a closed herd was deleterious and advocated in-and-in breeding as a good way of perpetuating a selected strain or set of characters. Providing that inbreeding depression did not appear, or undesirable recessives rise to an obvious level, this method was (and is) the simplest available to try to perpetuate a selected type. It requires the rigorous culling of off-types, but even so it is hard to avoid inbreeding depression after intense close breeding in a small population for three or four generations. This was the interval at which Southdown breeders generally 'took a cross'.[97]

Bakewell may have inbred intensely over a number of generations and he may have been lucky or have culled rigorously enough (or done both) to perpetuate his selected strain without degeneration. The procedure is chancy and most breeders, in practice, operate some line-breeding method to carry on with a desirable strain. Line breeding merely implies that in a closed flock or herd, several different blood lines, preferably from separate origins, and therefore genetically unrelated, are preserved in the herd and the lines regularly (or haphazardly) crossed in the matings at every generation to prevent the genetic relationship of the stock from rising too high. On an inter-herd basis this may involve constantly bringing in and crossing stock from other, noted herds. This requires more planning than incestuous inbreeding to type or individual, but the prerequisite of careful pedigree keeping was even more necessary here than it was for inbreeding. How much Bakewell, or others who recorded their matings carefully, may have kept their inbreeding indices low is now impossible to judge since so few detailed records have survived. Certainly in flocks and herds where individual matings were not recorded but a policy of home-bred replacement of stock was pursued (but with relatively little leeway to allow breedstock to be culled rigorously to select only the best), the use of carefully selected purchases or exchanges every three or four generations would have been a sensible way to avoid the undesirable aspects of inbreeding.

(f) *Performance testing.* Bakewell claimed that his selection of stock was for growth rate and food conversion characters, which could not have been assessed correctly unless they were measured. Without such a system of measurement, selection for performance was bound to be somewhat haphazard. It has already been noted how vague the assessment of milking performance in cattle was, although this was intrinsically easier to measure than either food conversion or the rate of live weight gain. There is abundant evidence that Bakewell, in his later years, made attempts to measure performance, although his procedures were unscientific and he never published the results of his trials. Most of these seem to have been comparative: to demonstrate the effects of crossing New Leicesters with other breeds or comparing New Leicester performance with that of other breeds. Young repeatedly pointed out the inherent defects in such procedures and it is probably true to say that Bakewell used the trials for demonstrations rather than for any genuine scientific purpose.[98] It seems unlikely that performance measurements were made in the foundation of the Dishley breed. It would have required a large staff and the presence of accurate weighing devices if it was to have been done properly. Systems for weighing stock alive in the stable or yard do not seem to have evolved until late in the eighteenth century, so Bakewell could not have used them in the formation of his breed.[99] Subjective assessments of performance were made in the selection of stock for breeding at Dishley in the later years, although the primary decisions on the stock which should be reserved for breeding were made on the basis of pedigree or blood line by that date.

(g) *Progeny testing.* In principle, the only sure way to measure the breeding potential of an animal is to test a sample of its progeny for their worth and the degree to which they have inherited the superior value of the parent. In practice, as has been seen, the interpretation of even the most sophisticated test is fraught with difficulty. Later in the century, all authors on the subject of the Dishley system were agreed that the hiring-out of sires by the season was used as a form of progeny test of the offspring of young tups from which their breeding value could be determined. The real point of ram hiring was, however, commercial: to obtain more money from each tup over several seasons than could be obtained by direct sale in any one season.[100] Furthermore, the system of hiring did not allow the hirer a very close examination of the trial get of his young rams. But it would seem that Bakewell, from a relatively early stage, did use the early offspring produced by a ram as a crude progeny test, withdrawing from service those rams which gave an exceptional performance and reserving them for his own use, so as to retain the lead in the furtherance of the Dishley stock. In the one well-documented instance where this occurred, his decision was based on lambs produced from his own ewes and not those which were produced as a result of letting.[101]

The later practice at Dishley and elsewhere under the aegis of the Tup Society had far more to do with theatre and the cunning exploitation of fashion than any relationship with the breeding value of stock. Ram prices were pushed higher by this combination or bidding ring than those commanded by the very best racehorses. Fashion and the reasonable desire to make a fortune fuelled the actions of the participants. The tup system became stratified, with the first-rank breeders trying to take over from Bakewell, while they in turn supplied second-rank breeders, whose function it was to supply working farmers at rates which it was economically worth their while to pay (usually less than ten guineas a tup). These latter prices remained at about the level that Bakewell and his predecessors had been able to charge earlier in the century and represented the real economic value of the stock to agriculture.[102] In the 1780s the point of breeding tups among the members of the Society had become totally divorced from any practical agricultural purpose and had become an end in itself. As such it could have no future once the bubble of fashion had burst.

(h) *The condition of stock.* In the general publicity for his breed, Bakewell stressed that the Dishley sheep were not only suitable for pasture grazing but would also survive well under hard conditions of climate and nutrition, a contention with which very few contemporaries agreed. In fact the breed was selected to respond well to very good conditions of nutrition and a great deal of care and attention in the general stockmanship. It could not possibly have performed satisfactorily as a folding sheep or as a hill breed, as the failure of Dishley half-breds on the hills constantly demonstrated. In practice, much of the phenotypic value of Bakewell's sheep was probably as much, or more, due to his high standards of husbandry as to the innate genetic worth of his animals. His whole farm was extremely productive since his arable husbandry was among the most advanced of his day. Many commented that his stocking rates appeared to be so high that each animal was not notably well fed, but these snapshot pictures of stock levels on individual weeks or days were not necessarily a good guide to the general stock levels on the farm. There can also be no doubt that the general standards of care and attention of all the stock at Dishley were very high indeed, especially in the later years when Bakewell had sufficient wealth to employ the staff to care for the animals really well. The buyers and hirers of Bakewell's stock were obviously far more influenced by their appearance than by performance or progeny test, since they could judge the first accurately for themselves, while the latter two they had to take more or less on trust. Modern stock breeders still reckon that show-ring quality, the production of an animal that looks first class, remains as much a matter of good nutrition and presentation as of inherent worth. Class can be put in 'by the mouth' and much of the reputation of Bakewell's stock certainly resulted from the superb standards of animal husbandry that he practised.

Whatever rumours may have developed about the original stock from which the Dishley was derived, there can be no doubt that the breed was a modified version of the lower Lincolnshire Wold sheep of the 1740s and 1750s. Commentators at the end of Bakewell's life and in the early nineteenth century found this hard to believe because the characteristic Wold sheep had entirely disappeared and Lincolnshire was occupied only by giant longwools and scruffy heath sheep on the remaining commons. Bakewell clearly stated the origin of his animals on more than one occasion and there really is no need to postulate any other breed in the formation of Bakewell's race.[103] Modern commentators have sometimes supported the theory of a mixed origin for the breed because they are impressed by the rigorous inbreeding which Bakewell said he practised. They have interpreted this as an example of the modern use of close inbreeding as a method of fixing the form of a hybrid following the cross of two separate strains or species. However, all the evidence suggests that inbreeding in the eighteenth century was employed as a means of retaining the excellence of selected individuals within a homogeneous closed herd, not as a method of fixing a heterogeneous hybrid.

Apart from Bakewell's own statements, there is the strong circumstantial evidence of his adoption of a career which derived from older activity on the Lincolnshire Wolds and probably in Teesvale as well. Bakewell took up the profession of tup or ram breeder, which was unusual in the East Midlands even in the 1790s, since most stock raisers in that region carried on their ancient tradition as graziers and were tup breeders as part of an overall grazing enterprise. However, throughout the eighteenth century there had been professional breeders of tups on the Lincolnshire Wolds, whose role was to supply tups to the marsh and lower-Wold graziers. These tupmen have evolved the practice of hiring their tups out by the season, almost certainly as a commercial proposition. They supplied rams to the graziers, who did not use their own home-bred rams as sires for reasons that are not now clear.[104] The graziers may have feared the consequences of inbreeding (although their large flocks made intense inbreeding hard to achieve) and always required an outcross of rams at each generation. But they could have adopted a scheme of swapping or buying in rams, as was done in Sussex, or of mutually exchanging rams at a modest fee, as was practised in Kent, so it is hard to see why specialist ram breeders should have arisen.[105] It seems

Plate 22. Robert Bakewell's Leicestershire ram, *Two Pounder*. Oil painting by J. Digby Curtis dated 1790. A portrait of one of Bakewell's leading rams from late in his career. It is not clear whether the painting was taken from life or not. Photograph from the Museum of English Rural Life, University of Reading.

unlikely that the system developed as a means of improving stock through ram selection or progeny testing on the part of the breeders. But it is evident that by the 1750s the system was being used as a means of improvement and that some of the ram breeders of Lincolnshire and Teesvale were beginning to develop reputations for quality and to charge for the hire of their rams accordingly. It would seem that Bakewell emulated their professional techniques exactly, the only difference between the Lincoln breeders and himself being the objectives for which the stock was bred. The Lincoln and Teesvale breeders selected for the classic feature of enlarged carcass size and for increased fleece weight as one method of countering the constantly falling price of worsted longwools throughout the eighteenth century. Their activities led to the production of the huge Lincoln and Teeswater longwools of the 1780s and 1790s.[106] The phenotype selected for was easy to judge and the direction of selection generally agreed, so that the effects of the fifty years of intense breeding in the marshes and rich grazings were spectacular – so spectacular that no one could credit the common origin of these great, lank marsh sheep with the smaller, blockier and faster-growing Bakewell Leicester. The reason for the difference was certainly the divergent selective pressures exerted by Bakewell for carcass quality, while the Lincolnshire breeders worked single-mindedly for fleece weight.

By 1745 the tup-breeding specialisation on the Lincolnshire Wolds was already well established. Thomas Staveley, of Kerton in Lincolnshire, hired tups from a local specialist breeder, whose general run of tups were selling for about £2 but who also possessed some tups valued at up to £30.[107] By the 1760s Thomas Pennant believed that in Lincolnshire tup breeders could sell rams, if they were very good ones (although he did not say what criteria were used to judge their value) at 50 guineas each and that hire fees of 20 guineas a season were not unusual. In the case of some high-grade rams, fees of 1 guinea per ewe were apparently obtained for service.[108] Tup breeding and hiring were also common in Teesvale at this period. William Storey, near Yarm in the vale, had a three-year-old tup in the late 1750s which weighed 398 lb (about 200 lb dead weight), which was sold to a Mr Banks of Harworth, near Darlington, for 14 guineas. This latter owner hired the ram out at the rate of half a guinea a ewe. By the mid-1760s the leading tup breeder in the vale was apparently Robert Gibson of Newsham Banks, who owned an offspring tup of the Yarm animal above, which was then hired at half a guinea a ewe, but which in the 1767 season he intended to hire at 1 guinea a ewe.[109] In the 1770s the large Lincolnshire graziers, the Dixon family, were beginning to hire out their own rams for a maximum of 3 guineas a season, but more usually at 1½ or 2 guineas. This suggests that as the Wolds were enclosed and the large marsh sheep spread onto the Wolds, the specialist tup breeders were disappearing and the graziers were hiring rams to each other on a mutual swap basis, much as seemed to be the practice in Kent.[110]

The hire fees obtainable among the leading Lincoln and Teesvale breeders of 20 guineas and upwards were higher than those which Bakewell often obtained at the same period. The fact that Lincoln rates were calculated per ewe suggests considerable care was taken over the mechanism of breeding and that it was done in closes or fields near the farm and reasonably closely supervised. By the 1780s the Dishley Society were keeping individual rams in small closes and only allowing small parcels of some half-a-dozen carefully selected ewes in to them at a time. Lincolnshire practice never became as sophisticated as that. From the 1750s through to the 1780s the Dixons selected parcels of ewes for each ram and separated each ram into a different close, but the animals were allowed to run loose together in numbers between 50 and 120.[111] Against this background of Lincoln tup-breeding activity, what can be said about the formation of Bakewell's breed from the same general source in the 1740s and 1750s?

Plate 23. Hiring out rams at Dishley, Leicestershire. Oil painting by Thomas Weaver, 1810. The circumstances of this painting are unrecorded. It was not contemporary with the famous Bakewell hirings in the late 1780s and early 1790s. It may represent events at Dishley in the early nineteenth century when the farm was run by Bakewell's nephew, Robert Honeyborn, who carried on the trade of specialised stock breeding. Photograph from the Tate Gallery, London.

Bakewell was not apparently the first Leicestershire breeder to seek stock from the Lincolnshire Wolds. Joseph Allom of Clifton and possibly Stone of Godeby, near Melton Mowbray, obtained stock from this source at some time in the first half of the eighteenth century. Allom seems to have adopted the profession of ram breeder and at Clifton, Leicestershire farmers could obtain rams each season for 2 or 3 guineas each. Allom was selling, not hiring, rams as many Lincoln tupmen still were in the early eighteenth century.[112] There may well have been others who also specialised in the trade, so that Bakewell may have had local models of this profession as well as experience gained on trips to Lincolnshire. Whether Bakewell used Stone or Allom for tups in the creation of the Dishley breed has not been recorded. The habit of buying rams at specialist autumn shows seems to have been an institution of considerable antiquity in Leicestershire, since the Leicester ram fair was already ancient when the hiring of rams was introduced there in the 1780s and 1790s.[113] Bakewell seems to have spotted the commercial benefits of hiring rams, rather than selling them, at an early date and to have been instrumental in introducing this technique into Leicester at some time in the 1740s, at the beginning of his career as a sheep breeder. Marshall spoke vaguely of a date around 1740 and prices of 16s and 17s 6d for the hire of rams for the season (not per ewe as some commentators have supposed). Young believed the year in question was 1744. Marshall thought that Bakewell's ram prices had remained low until the 1780s. In 1780 Bakewell was letting rams at only 10 guineas each for the season, while one Parkes, at Quarndon, let one tup in the same year for 15 guineas, although this animal was itself of Dishley stock.[114] However, there can be little doubt that Bakewell had achieved better prices than these at an earlier date and that the 1780 rates reflect the fact that Bakewell was recovering from bankruptcy and a period when the farm administration had been out of his hands (in the mid-1770s).[115] At least one tup had been hired at 40 guineas for the season in the mid-1760s and in 1774 the same price had been obtained for the service of a ram to cover 20 ewes.[116] In general, Bakewell's stock was probably available at similar prices to those obtained by the best tup breeders in Lincolnshire in the 1750s and 1760s. Publicity, fashion and guile began to produce their most spectacular effects only in the 1780s.

The precise source of Bakewell's Lincolnshire Wold sheep remains vague. A respondent of Arthur Young's in the early nineteenth century believed sheep had been obtained from Sleaford, from near Grantham and from Melton Mowbray, where other breeders may well have been importing Lincoln sheep in the 1740s. Lincoln sires were probably being used on Leicestershire sheep as far back as the beginning of the eighteenth century. Another strongly rumoured source was the region around Bilsby and perhaps a Lincoln ram breeder of that area known as Mr Stow. George Culley thought that this Stow was resident at Long Broughton and had procured many Lincoln tups for Bakewell. If this Broughton was the one in South Humberside, in the very north of the county, it may have given rise to a later story that Bakewell had drawn his base stock from the Yorkshire Wolds.[117] It is certainly probable that Bakewell bought rams from a number of leading Lincoln tup breeders and that he worked on a ewe population drawn from the local Leicestershire pasture sheep which were already of the same general type. His original stock was therefore diverse, but all of one breed form.

The only reasonably reliable source which can tell us anything about Bakewell's early career is his disciple, George Culley, who knew him and hired rams from him from the early 1760s. Culley had stayed at Dishley in 1762 as part of a tour to learn his trade of farming, a practice which seems to have been conventional among the sons of the better-off and more progressive farming families. Back in Durham, George and his brother,

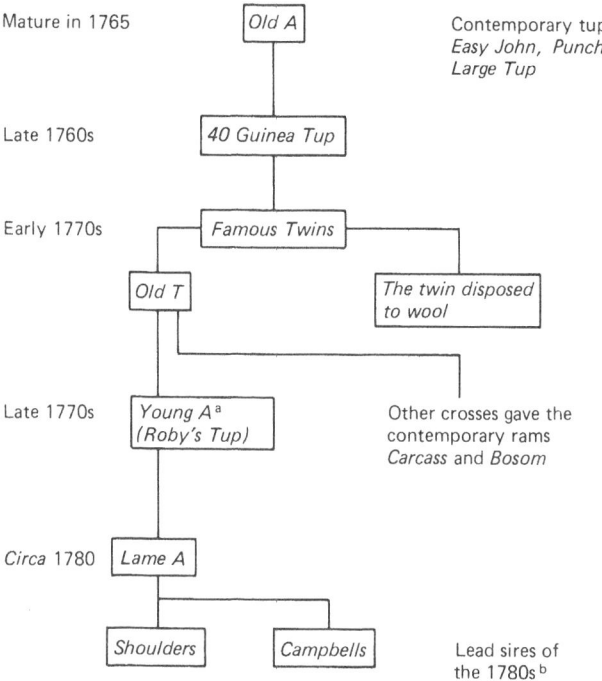

Chart 3. Outline pedigree of Bakewell's main sire line

Matthew, determined to hire Bakewell rams and grade up the sheep of their farm and of the Fenton region of Northumberland, where they had moved in 1767.[118]

George hired his first Bakewell ram in 1763 or 1764 and systematically attempted throughout his life to develop a secondary centre of pure Dishley stock in the northern counties. From the account given by Culley of the original ancestry of his own stock it is possible to construct a basic picture of some of Bakewell's sheep in the early 1760s.[119] The first two tups that the

---

[a] *Young A* shared lead ram status with *Charles*. A progeny test on the farm at Dishley showed that *Young A* produced superior offspring. John Breedon, who had been Bakewell's shepherd since 1761 and had therefore supervised the whole breeding programme outlined here, persuaded Ashworth not to sell him as the latter had originally intended. See J.H., 'Visit to Bakewell's farm'. The other shepherd on the farm in 1793, William Arnold, had only been there ten years.

[b] G. Culley, *Description of sheep in the possession of Messrs Culley of Northumberland. In answer to queries circulated by Sir John Sinclair* (1792), p. 93; Rowe, 'The Culleys', p. 162.

Culleys hired were called *Punch* and the *Large Tup*. Either in the following year or possibly the year after (1765 or 1766) they hired another ram known as *Easy John* or *Sober John*. This was not Bakewell's best ram, which at this time was called *A*, and subsequently *Old A* to distinguish it from one of its second-generation progeny also known as *A*. The Culleys did not hire this ram because Bakewell was asking more than they could (or would) pay for it. *Old A* seems to have occupied the same position in the structure of Bakewell's sheep flock as the bull, *Twopenny*, did in his cattle herd: the first animal that he had bred which had achieved his major objective in carcass conformation. Thus he increased the hire charge for *A* and apparently adopted the formal system of giving his top sires letters rather than names, a practice which he later adopted for both cattle and horses. Bakewell's leading rams in the 1760s and 1770s were all derived from this sire, *Old A*, the first-generation offspring being the tup, *40 Guineas* (according to Culley the only one to have achieved this price before the early 1780s), while this tup, in turn, sired the pair known as the *Famous Twins*, one of which was a fine carcass animal, called *Old T*, while the other was noted for its wool and was therefore of less interest to Bakewell. *Old T* produced a ram offspring in the third-generation known as *Young A* or *Roby's Tup*, the lead sire at Dishley in the mid-1770s. This tup was nearly disposed of to a Mr Roby by the administrator at Dishley during Bakewell's bankruptcy, one Ashworth. *Young A* was in turn the sire of *Lame A*, which was lent to Culley free of charge in the late 1770s or early 1780s in recompense for the financial assistance which he had given Bakewell during the bankruptcy. This same sire, *Lame A*, produced two rams at Dishley, known as *Shoulders* and *Campbells*, his most noted rams at the beginning of the fashionable explosion in Dishley prices. Thus, there seems to have been an unbroken sire line pedigree from the original tup which showed the major positive characters that Bakewell sought, to the lead tups at the period of high fashion.

According to Culley, Bakewell did not import any new material into his flock after the mid-1760s, so the ewes involved in this programme could all have been reared at Dishley from sheep previously assembled in the period between the mid-1740s and the mid-1760s.[120] There is no extant evidence of intense inbreeding during the period covered by this pedigree – from 1765 to the 1780s. Rams of other blood lines were probably present, for instance the alternative lead sire, *Charles*, in the late 1770s and *Easy John* in the 1760s, and there is no reason to suppose that the ewes employed in the lead sire line were of the same origin or that close sibling – sibling or parent–sibling inbreeding was taking place. It is more likely that Bakewell believed, with so many of his contemporaries, that sire line inheritance was all that really mattered and that by rigorous culling of the rams which he did not favour he was able to perpetuate the desirable carcass type which he seems to have hit upon first in *Old A*. If he had been inbreeding intensely during the formative years of the Dishley strain, his stock would almost certainly have suffered from inbreeding depression. From the mid-1780s onwards he may have begun to inbreed his stock heavily in an attempt to cut down on the number of off-types and to increase the frequency of desirable rams, and this may explain the declining fecundity of Dishley stock at this late stage.[121] But there is as yet no evidence for intense inbreeding in the formation of the Dishley race.[122]

It cannot be without significance that the names he gave to some of his rams, such as *Shoulders, Bosom, Carcass, Campbells* (hocks) etc., applied to certain conformation areas where the animals concerned excelled in Bakewell's opinion. It lends support to the contention that Bakewell's main objective in selection was to improve the form of the carcass and its readiness to fatten, choosing his stock by the traditional 'feel' of the Midland grazier. In fact, one could argue that his choice of meat rather than wool as an object of selection was governed solely by his background in a grazing area where the carcass had traditionally been the main object of interest. He was certainly correct to see that early and rapid growth and efficient conversion of food into flesh were important, but for reasons already rehearsed it seems unlikely that these could have been realistic objectives for selection or in the formation of the Dishley breed. Bakewell's deliberate selection policy was based entirely on appearance, with a bonus of early maturity resulting from his desire to bring lambs on as fast as possible to reproductive age, while no serious attention whatsoever seems to have been paid to food conversion. It remains doubtful whether the mature breed was significantly more efficient at converting food into carcass than any other breed. Comparative trials between breeds for such physiological characters are notoriously difficult to conduct and hard to interpret correctly. Contemporary trials were suspect in both design and execution and their results must be treated with caution. However, several such trials were described in the *Annals of Agriculture* in the 1790s, of which that conducted by the Duke of Bedford was the only one of any real merit.[123]

Lambs of four breeds, Southdown, Leicester, Worcester and Wiltshire, were compared in groups of 20 (16 for the Wiltshires) from November 1794 to February 1796, when they were killed at Smithfield. Only the turnips and hay fed to them could be weighed, on pasture the amounts eaten could not be recorded. The amounts of turnips and hay eaten by all four groups were very similar and it must be assumed the sheep all consumed similar quantities of food at pasture. The comparative weight gains were as follows:

Southdowns	40.05 lb per head
Leicesters	40.75 lb per head

Worcesters		50.40 lb per head
Wiltshires*		52.40 lb per head

* However, since there were only 16 sheep they were eating more per head. If there had been 20 animals they would have put on some 20% less per head: approximately 41.90 lb per head.

Converting these figures into percentage weight increases so that the larger size of the Worcesters and Wiltshires may be discounted, gives the following results:

Southdowns		41.5% increase in weight
Leicesters		42.0% increase in weight
Worcesters		54.0% increase in weight
Wiltshires*		48.0% increase in weight

* Reckoning that these animals would have given 20% less if the sample size had been correct, the percentage increase here would have been about 40%.

These figures, for what they are worth, give no support to the view that either of the improved breeds were significantly better converters of food into flesh; in fact the Worcesters seem to have been considerably better performers.[124] Reports in the *Annals* also call into question the reality of the supposed improvements in the ratio of carcass meat to offal in the new breeds. Some individual assessments of the carcass ratio, from various authors, are summarised below:

Breed	Carcass weight as percentage of live weight[125]			
Southdown	60.5	60.6	55.0	59.3
Leicester	59.6	59.9	65.4	
Romney Marsh	66.9	64.8		
Worcester	55.3			
Cotswold	59.1			
Wiltshire	55.4	60.0		
Norfolk	56.3			

The Romney Marsh and the Cotswold seem to be as good, if not better, than the improved breeds. All this goes to reinforce the opinion expressed here that whatever the sophisticated objectives Bakewell claimed he was breeding for, in practice the basis of the difference between the Dishley and other breeds lay entirely in conformation and general appearance and perhaps its superior growth rate. It must be doubtful if food conversion or carcass ratio were significantly improved, or that the fundamental form of the carcass was changed either in the Bakewell strain of Longhorn cow or in Lincolnshire Wold sheep. However, the animals *looked* much better grazier's animals, with their tendency to fat up and round out well. In a sense, all Bakewell had done was to create a new, if somewhat more rational fancy for sheep of a particular shape, rather than merely tinkering with colour or horn form. On the other hand, it would seem that the Lincoln breeders had genuinely succeeded in breeding animals with a greatly increased fleece yield, although the death of the longwool market made their achievement a pointless one. Sensibly they reverted back towards the form of the Wold sheep from which they had started, of which the best surviving examples were the Dishleys. The use of Dishley stock by the Lincoln breeders was, of course, made much of by the Leicestershire men, although it probably did not have the significance the latter ascribed to it. Certainly Bakewell must take considerable credit for publicising the idea of selecting stock for economic performance, but whether his actual achievements in this field were of any significance remains very doubtful.

8

◆◆

Summary and conclusion

Two basic questions were asked at the beginning of this book: was there any measurable increase in the productivity of commercially bred livestock and, if there was, to what major causes could any such change be attributed? As a result of the investigations pursued here, some tentative conclusions may be drawn.

There can be little doubt that the overall size of the average individual sheep and draught horse increased from the sixteenth to the eighteenth centuries and that the average fleece weight of the sheep also went up. The size and yield data for cattle are not sufficiently copious to provide much basis for the assessment of any increase in average size although, as with sheep and horses, evidence for widespread breed size variation is strong throughout this period. Considerable archaeological evidence exists that there was a dramatic increase in the size of cattle in England between the thirteenth and early sixteenth centuries. Up until the thirteenth century the size appears to have been similar to that found from the Romano-British period, i.e. very small cattle, while a size increase and the replacement of short-horned by long-horned animals occurred in the fourteenth and fifteenth centuries. The cattle of the late seventeenth and early eighteenth centuries appear to have been comparable in size with modern animals. The author of the latest review of this subject proposes that improvements in nutrition, rather than positive livestock selection or replacement by foreign breeds, were most likely to have been instrumental in generating this transition during the fourteenth and fifteenth centuries. From the evidence of the breeding advice offered in the seventeenth and eighteenth centuries such an accidental, domestic–environmental selection for larger size does seem the most likely explanation.[1] Average milk yields may also have increased in the seventeenth and eighteenth centuries but there is as yet insufficient data to make any firm judgement. There is a high probability that the sprinting Thoroughbred horse was a faster racing animal than its contemporaries in the mid-eighteenth century, but it must be doubted if racing performance has improved since then.

In the ox and the sheep the major criteria for judging biological efficiency are such characters as yield per unit of food input and end-product quality. In the case of the horse, these productive characters are less important compared with aspects of physiological performance, such as weight-drawing or carrying capacity, aesthetic action, speed, endurance and courage. Since the qualities looked for in horses more nearly approach those thought desirable in humans it has been argued that anthropomorphic thinking dominated the views of horse breeders from ancient Greek times onwards. For breeders of both agricultural and racing stock, however, there were other features generally classed as fancy points which were considered desirable but whose relationship with productivity or quality has always been extremely tenuous. Horn form, coat colour, tail length, colour marks, beauty rather than usefulness of conformation, have always fascinated breeders and continue to do so today, when agriculture has supposedly become a wholly rational and unemotional exercise. This position was reinforced by classical concepts of aesthetics, which implied that what looked beautiful possessed virtue, that external beauty was linked to internal character. The concept could presumably be applied to productive agricultural characters as well as points of personality. Savants such as Houghton in the late seventeenth century were rather scornful of the fancy and felt some rational analysis of productive character was required. This reasoned approach to productivity, displacing the older aesthetic principles, seemed only to spread slowly in the eighteenth century.

Since fancy characters generally have far simpler genetic control than complex productive or physiological properties, they respond far more quickly, positively and surely to selection by eye, which has ever been the hallmark of the breed fancier. It is not surprising, therefore, that all breeders have made great play of aesthetic characters. They were the only ones where simple selection led to predictable results. Such fancy characters also satisfied the emotional commitment which all breeders must feel for their material if they are to be successful in a difficult and chancy enterprise where long perseverance rather than quick results are the order of the day.

The breeders were perhaps wise to stick with those characters which seemed to be firmly hereditary and responsive to simple positive selection. All productive and physiological characters are not only complex in their inheritance but their expression is heavily influenced by interaction with the environment. It is small wonder that many authors could not persuade themselves until the latter half of the seventeenth century that there were any such truly heritable or innate aspects of these characters which could be selected with a reasonable chance of improvement. It is not without significance that the first successful attempt (in twentieth-century terms) to formulate hereditary principles – by Mendel – was achieved with essentially fancy rather than horticulturally useful features of the pea plant and that the genetics of productive characters have proved extraordinarily difficult to work out. The science of quantitative genetics employed to analyse these characters today is by far the most difficult branch of the subject to grasp. Since even today many commercial breeders, as opposed to academic geneticists, have only an elementary understanding of the simplest Mendelian principles, it is not perhaps surprising that successful improvements in productive characters as a consequence of deliberate breeding policy have proved to be so rare in the past.

In productive characters, three major criteria and their interaction with each other must be considered: the size, volume or weight of the product aimed at, the rate of growth of the product and the efficiency with which an animal converts food input into product output. Only the first of these is susceptible to easy measurement, so that evidence of a drive to increase the size of animals and the weight of fleece in the sheep has been reasonably obvious from at least the seventeenth century. To what extent the aim of improvement in size and weight was achieved by selective breeding remains unclear, but most evidence points to its fulfilment through breed replacement, or to a lesser extent through grading-up, rather than by positive selection for size increase among variants in a local population. In turn, the most likely explanation for the existence of size variants among domestic breeds in the first place is as a consequence of random and natural selection within different domestic environments rather than any earlier deliberate breeding policies in animal husbandry. Positive selection within a local population to produce a significant increase in carcass size and fleece weight in sheep may have been successful on the Lincolnshire marshes and perhaps in the East Midlands in the eighteenth century, but most other examples of higher yield and larger size in local areas can best be ascribed to breed replacement or grading-up. The incentive to change to a bigger breed seems to have been the desire to improve the overall productivity of arable and pastoral husbandry through better agrarian management.

A more rapid growth rate, and especially the early onset of sexual maturity, are probably accidental consequences of the very processes of domestication and controlled animal breeding. The practice of Lancashire and, later, Midland Longhorn breeders appears to have been a peculiarly specific example of the same phenomenon at work, in this case a selection for vigour and 'covering power' in young bulls, probably accentuating the genetic tendency towards early maturity in all members of the breed. However, there is little evidence before the late eighteenth century that early maturity was deliberately selected for as part of breeding policy.

Food conversion efficiency was probably a well-understood concept in the late seventeenth and early eighteenth centuries, but there is no evidence of any attempt to select stock with that character in mind before the activities of Bakewell later in his career. It has been argued, however, that despite his awareness of the critical role of this character in the assessment of livestock productivity, he had neither the measuring equipment nor the genetic understanding necessary to select for this property. However, it was the realisation of the importance of this character and the desirability of its control which was a major part of Bakewell's contribution to improved animal husbandry. There can be no doubt that the techniques that he recommended, if not the livestock which he himself bred, had an enormous influence on all nineteenth-century breeders.

The horse breeders of the sixteenth century were already aware of the hereditary nature of size in their stock and, as in sheep and perhaps cattle breeding, it is likely that some selection among local variants to increase breed size could have been attempted. Whether anything like this occurred in England seems doubtful, but from various hints of the intense interest taken in horse breeding in the Low Countries it is possible that the increase in size of Flemish horses in the seventeenth century was the result of successful positive breeding in this direction. However, the major concerns of horse breeders were more nebulous characters such as courage, endurance and action. As would be supposed, they seemed to have been able to make little progress in this direction by selection among local variants. A policy of breed replacement and grading-up emerged among horse breeders in the late seventeenth century from the chaotic experience of multiracial cross-breeding which had been indulged in in the sixteenth century. To pick some kind of pathway through this massive mongrelisation, akin perhaps to the Roman experience with dogs, sixteenth- and seventeenth-century breeders attempted to impose their own controlled hybridisations between stock known or supposed to be of a specific type. Since imports were those most sure to be 'pure' in the sense of unmixed in the local mongrel collection, the significance of import sires became very great in breeding policy. The rising fashion for horse racing in the late seventeenth and early eighteenth centuries encouraged some breeders to try to continue 'pure' Oriental blood horses in England, initially to provide favoured sires for the production of racehorses by hybridisation with local mares. This was probably the origin of the 'Thoroughbred'. This local Oriental variant, modulated with North European genes, eventually itself produced a sprinting racehorse of a type superior to either the traditional cross-bred racer or Galloway saddle horses. Thus, by the mid-eighteenth century, the Thoroughbred was fast becoming the major racehorse breed, a position consolidated by the appearance of the *General Stud Book* and the closing of the breed to any further influences.

Purity of breed was the main objective – to overcome the general mongrelisation and to produce a native 'Oriental' type. The mechanism for ensuring this purity, the accumulation of pedigree data, then spread to the agricultural breeders of the eighteenth century. Bakewell certainly maintained very careful records of his own sheep and cattle and the breeders of high-class draught horses adopted the same system. Inbreeding had been advocated in horse breeding in the seventeenth century, but only in the context of a programme of grading up towards imports, not in the framework of a small, closed breed. Bakewell expanded the concept by his recommendation of inbreeding in small, closed herds and flocks, although it is doubtful whether he himself created his own breeds by using any such process. Marshall spoke of inbreeding as common at Newmarket in the 1790s, and suggested that Bakewell derived his views from there.[2] However, it seems more likely that racehorse breeders were experimenting with a procedure then being discussed among fashionable agricultural breeders. Certainly the only known case of a breed being created by intense inbreeding occurred in the early nineteenth century with the foundation of the modern Shorthorn cattle breed, said to have been based directly upon Bakewell's teaching.

However, Bakewell's most significant idea was probably the enunciation of the principle of the progeny test to arrive at a judgement of parental breeding value. Hints of such a practice appeared in antiquity, but there is no record of the concept in the early modern period before Bakewell's time. But, as with the measurement of the efficiency of food conversion, Bakewell had not the necessary techniques to conduct anything other than a rather crude and misleading test. In practice, he probably relied as much upon pedigree for an assessment of breeding value. Here he was almost certainly borrowing from the practice of Thoroughbred breeders. So successful had the Thoroughbred become as a racehorse by the mid-eighteenth century, a success founded upon the belief that heredity was far more important than development or environment in character expression, that by the time that Osmer and Wall were writing, a complex folklore of 'blood' lineages had evolved. It is highly probable that Bakewell borrowed both the pedigree concept and its use as a basis for breeding value judgement from the racehorse breeders. But since other ideas seem to have flowed in the reverse direction, some interactive model of horse and agricultural breeding technologies is probably required to understand the relationships between these two most important groups of eighteenth-century animal breeders properly.

Notes

The following abbreviations have been used throughout the notes and bibliography.

AHR	*Agricultural History Review*
APC	Acts of the Privy Council
CLPF & D	Calendar of Letters and Papers, Foreign and Domestic
CSPD	Calendar of State Papers, Domestic
EHR	*Economic History Review*
ESRO	East Sussex Record Office
HFRC	Historic Farm Records Collection, University of Reading
HMC	Historical Manuscripts Commission
KAO	Kent Archive Office
LAO	Lincoln Archive Office
RO	Record Office
WSRO	West Sussex Record Office

Preface

1 Nicholas C. Russell, 'Animal breeding in England *c.* 1500–1770', unpublished Ph.D. thesis, University of London, 1981.
2 e.g. M. Abercrombie, C.J. Hickman and M.L. Johnson, *A dictionary of biology* (Penguin, 1951 and subsequent editions).

Introduction

1 Modified after F.E. Zeuner, *A history of domesticated animals* (Hutchinson, 1963), p. 63.
2 R.J. Berry, 'The genetical implications of domestication in animals', in P.J. Ucko and G.W. Dimbleby (eds.), *The domestication and exploitation of plants and animals* (Duckworth, 1969), pp. 207–18.
3 Berry, 'Genetical implications', p. 212.
4 I.L. Mason, 'The role of natural and artificial selection in the origin of breeds of farm animals', *Z. Tierzuchtg. Zuchsbiologie.* 90 (1973), 229–44.

5 C.R. Darwin, *The variation of animals and plants under domestication* (John Murray, 1868). The role of unconscious domestic selection as a model in the theory of evolution is discussed in P.J. Vormizzer, *Charles Darwin: the years of controversy* (University of London Press, 1972), pp. 170–4.
6 J.P. Scott, 'Evolution and domestication of the dog', *Evolutionary Biology*, 2 (1968), 244.
7 Berry, 'Genetical implications', pp. 209–11; Mason, 'Natural and artificial selection', pp. 231, 240.
8 Berry, 'Genetical implications', pp. 211–13.
9 Berry, 'Genetical implications', quoting from H. Spurway, 'Can wild animals be kept in captivity?', *New Biology*, 13 (1952), 11–30.
10 A. Loudon and J. Fletcher, 'Monarch of the farm', *New Scientist*, 99 (14 July 1983), 88–92.

1 Breeding strategies

1 Discussion of present-day strategies of livestock breeding and their operation can be found in J.E. Nichols, *Livestock improvement in relation to heredity and environment* (Oliver and Boyd, 1957); A.L. Hagedoorn, *Animal breeding* (Crosby Lockwood, 1962); I.M. Lerner and H.P. Donald, *Modern developments in animal breeding* (Academic Press, 1966); I. Johanssen and J. Rendel, *Genetics and animal breeding* (Oliver and Boyd, 1968); Sir J. Hammond, *Hammond's farm animals* (Edward Arnold, 1971); and D.S. Falconer, *Introduction to quantitative genetics* (Oliver and Boyd, 1980).
2 For example, the sheep breeders of the Shetland Islands in the late eighteenth century adopted such a strategy, although natural selection evidently buffered its effects. See John Tulloch, 'Abstract of an account of Shetland sheep', an appendix to Andrew Ker, *Report . . . of the state of sheep farming along the eastern coast of Scotland, and the interior parts of the Highlands* (Edinburgh, 1791), p. 43.
3 For instance, in classical times, various passages from Aristotle clearly demonstrate this belief. See *Historia animalium*, ed. W.W. D'Arcy Thompson (Oxford University Press, 1910), book VIII, 605^b and 606^b, book III, 519^a.
4 This was John Ray's view in the seventeenth century. See 'Mr Ray on the number of plants' (undated), in W. Derham (ed.), *Philosophical letters between the late learned Mr Ray and several of his ingenious correspondents . . .* (London, 1718), pp. 344–7.
5 J. Ray, 'Mr Ray on the number of plants'.
6 Nichols, *Livestock improvement*, pp. 177–80; W.J. Carlyle, 'The changing distribution of breeds of sheep in Scotland, 1795–1965', *AHR*, 27 (1979), 19–29.
7 Nichols, *Livestock improvement*, chapter 14, *passim*; C.E. Terrill, '50 years progress in sheep breeding', *Journal of Animal Science*, 17 (1958), 944–59.
8 C.M. Prior, *The history of the racing calendar and stud book* (Horse and Hound, 1926), p. 11.
9 For an account of the three traditions within seventeenth- and eighteenth-century science, see, for instance, H. Kearney, *Science and change, 1500–1700* (Weidenfeld and Nicolson, 1971).
10 For instance, John Aubrey was a good example of this view. See O.L. Dick (ed.), *Aubrey's brief lives* (London, 1972), pp. 82–93.
11 Jacob, *The logic of living systems*, trans. B.E. Spillman (Allen Lane, 1974), p. 1.

2 The classical tradition: theories of heredity and breeding practice in Greece and Rome

1 H. Stubbe, *History of genetics*, trans. T.R.W. Waters (MIT Press, 1972), pp. 17–20.
2 G.E.R. Lloyd (ed.), *Hippocratic writings* (Penguin, 1978), pp. 317–20.
3 Lloyd, *Hippocratic writings*, pp. 320–1.
4 Lloyd, *Hippocratic writings*, p. 323.
5 Lloyd, *Hippocratic writings*, pp. 326–7.
6 Lloyd, *Hippocratic writings*, p. 162.
7 Stubbe, *Genetics*, pp. 33–41.
8 Aristotle, *De generatione animalium*, ed. A.L. Peck (Loeb Classical Library, 1943), book I, p. 97.
9 A.L. Peck, Introduction to *De generatione*, p. lxv.
10 Aristotle was adopting the much older view that semen was drawn from blood and trying to provide rational justification for it. See C. Zirkle, 'The early history of the idea of inheritance of acquired characters and of pangenesis', *Trans. American Phil. Soc.*, new series, 35 (1946), 121.
11 Aristotle, *De generatione*, book I, p. 91.
12 Aristotle, *De generatione*, book I, p. 117.
13 Aristotle, *De generatione*, book I, p. 103.
14 Stubbe, *Genetics*, p. 39.
15 A.L. Peck, Introduction to *De generatione*, pp. xi–xv.
16 A.L. Peck, Introduction to *De generatione*, p. lxxii.
17 Aristotle, *De generatione*, book IV, pp. 395–9; Aristotle, *Historia animalium*, book VI, pp. 573b–574a.
18 Aristotle, *De generatione*, book IV, p. 411.
19 Aristotle, *De generatione*, book IV, pp. 395–417, *passim*.
20 Aristotle, *De generatione*, book II, *passim*.
21 Stubbe, *Genetics*, pp. 41, 45–7.
22 Xenophon, *On hunting*, in *Scripta minora*, ed. E.C. Marchant and G.W. Bowersock (Loeb Classical Library, 1968), p. 415.
23 Plato, *The republic*, ed. D.P. Lee (Penguin, 1955), pp. 213–15.
24 Xenophon, *Constitution of the Lacedaemonians*, in *Scripta minora*, pp. 137–41.
25 Aristotle, *Historia animalium*, book V, p. 516h.
26 Aristotle, *Historia animalium*, book V, p. 546a and book VI, pp. 575a–575b.
27 Aristotle, *Historia animalium*, book VI, p. 576a.
28 Aristotle, *Historia animalium*, book IX, p. 631a.
29 Aristotle, *Historia animalium*, book V, pp. 545a–546b, *passim* and book VI, *passim*.
30 Aristotle, *Historia animalium*, ed. A.L. Peck (Loeb Classical Library, 1965), book III, pp. 228–9.
31 A.H. Broderick (ed.), *Animals in archaeology* (Barrie and Jenkins, 1972), pp. 105–6.
32 Aristotle, *Historia animalium*, ed. Thompson, book III, p. 523a; Pliny, *Natural history*, book XXV, quoted in K.D. White, *Roman farming* (Thames and Hudson, 1970), p. 277.
33 Aristotle, *Historia animalium*, book VI, p. 572a.
34 Aristotle, *Historia animalium*, book VI, p. 575a.
35 M.T. Varro, *Res rusticae*, ed. W.D. Hooper and H.B. Ash (Loeb Classical Library, 1960).

36 Varro, *Res rusticae*, book II, p. 311.
37 Varro, *Res rusticae*, book II, p. 373.
38 Varro, *Res rusticae*, book II, p. 339.
39 Varro, *Res rusticae*, book II, p. 375.
40 Varro, *Res rusticae*, book II, p. 371.
41 Varro, *Res rusticae*, book II, p. 373.
42 Varro, *Res rusticae*, book II, p. 345.
43 Varro, *Res rusticae*, book II, p. 397.
44 Varro, *Res rusticae*, book II, p. 329.
45 Varro, *Res rusticae*, book II, p. 363.
46 Varro, *Res rusticae*, book II, p. 403. The later writer, Columella, by contrast, thought that all pups should be raised, if the bitch was finding it difficult, by feeding them on goat's milk. See Columella, *De re rustica*, ed. E.S. Forster and E.H. Heffner (Loeb Classical Library, 1954), book VII, para. XII. Pliny also suggested a number of basic criteria by which a proportion only of a litter should be raised. See Pliny, *Natural history*, ed. H. Rackham (Loeb Classical Library, 1961), vol. 3, book VIII, para. LXII. Merlen reports another method of selection by culling the litter but does not give its source. See R.H. Merlen, *De canibus: dog and hound in antiquity* (J.A. Allen, 1971), p. 66.
47 Varro, *Res rusticae*, book II, pp. 333, 343, 347, 353, 359.
48 Varro, *Res rusticae*, book II, p. 387.
49 Varro, *Res rusticae*, book II, p. 390.
50 White, *Roman farming*, pp. 26–8.
51 Columella, *De re rustica*, book VI, paras. XVII, XXXVII.
52 Columella, *De re rustica*, book VI, para. I.
53 Columella, *De re rustica*, book VII, paras. II, III.
54 Columella, *De re rustica*, book VII, paras. IX and III.
55 Columella, *De re rustica*, book VII, para. XXXVI.
56 This notion had originated with Aristotle. See *Historia animalium*, book VI, p. 574[a].
57 Columella, *De re rustica*, book VI, para. XXXVII.
58 Columella, *De re rustica*, book VI, para. XXXVII.
59 Columella, *De re rustica*, book VII, para. II.

3 Generation and the market: the background to animal breeding in the seventeenth and eighteenth centuries

1 E. Gasking, *Investigations into generation, 1651–1828* (Hutchinson, 1967), pp. 7–10.
2 J. Thirsk, *Horses in early modern England: for service, for pleasure, for power* (University of Reading, 1978), pp. 20–1.
3 Canon Scott-Robinson, 'The expense book of James Master Esq., 1646–1676', *Arch. Cantania*, 15 (1883), 16 (1886), 17 (1887) and 18 (1889).
4 For instance, Andrew Snape, farrier to Charles II, published a general text on horses as the *Anatomy of an horse* in 1683. The work contained an appendix on 'Generation of animals'. This concentrated exclusively on a summary of recent anatomical work.
5 Gasking, *Investigations*, chapter 2.
6 Harvey made observations on pregnancy in deer. Both fallow and red deer have abnormal early foetal development compared with other mammals, being small

and hard to see. Harvey was notoriously conservative in the use of lenses or microscopes to assist in the examination of small objects. See A.W. Mayer, *An analysis of the 'De Generatione Animalium' of William Harvey* (Stanford University Press, 1936), pp. 103–20; and G. Keynes, *The life of William Harvey* (Oxford University Press, 1978), pp. 344–7, 339–42.

7 W. Harvey, *Anatomical exercitations . . .* (London, 1653), for instance pp. 167–8, 269. In some places he went so far as to regard the seminal contributions as resembling a contagion (pp. 260–1).

8 Harvey's appeal to God to direct events was one of the last serious attempts to do so before the domination of mechanical philosophy. Once Descartes had denied that animals other than man required rational souls, the way was open for a purely mechanical interpretation of biological generation. See Jacob, *Logic of living systems*, pp. 22–3.

9 Both Mayer and F.J. Cole (*Early theories of sexual generation* (Clarendon Press, 1930)) take this view, although Gasking disagrees.

10 He was also fairly hard on his old teacher at Padua, Fabricius. See Keynes, *Life of William Harvey*, pp. 347–8.

11 Sir K. Digby, *Two treatises, in one of which the nature of bodies, in the other the nature of man's soule, is looked into, in way of discovery of the immortality of reasonable souls* (Paris, 1644), p. 227. Reactionary critics, such as Alexander Ross, laid into all the leading natural philosophers of their day. The critique by Ross of Harvey's feeble position on the ethereal male reproductive contribution was entirely reasonable. See Alexander Ross, *Arcana microcosmi or the hid secrets of man's body discovered* (London, 1651), pp. 224–35.

12 Gasking, *Investigations*, chapter 3.

13 P.J. Bowler, 'Preformation and pre-existence in the seventeenth century: a brief analysis', *Journal of the History of Biology*, 4 (1971), 221–44.

14 Bowler, 'Preformation and pre-existence', pp. 222–7.

15 Jacob, *Logic of living systems*, pp. 60–3; Gasking, *Investigations*, chapter 3.

16 Bowler, 'Preformation and pre-existence'.

17 Linnaeus for instance dismissed all subspecific variation as derived 'only from accidental changes, generally owing to the climate, soil, heat, winds etc.' (quoted in Jacob, *Logic of living systems*, p. 48). In the eighteenth century Cuvier among others extended this by arguing that the superficial variations of organisms were trivial and allowed to vary since they were not retained under central control. See Jacob, *Logic of living systems*, pp. 153–4.

18 Jacob stresses the relationship between the divine plan classification, mechanical philosophy and *emboîtement*, especially in the eighteenth century. So powerful was this mutually reinforcing triumvirate that physiological or hereditary experiments which were conducted were interpreted exclusively as support for *emboîtement*. See Jacob, *Logic of living systems*, pp. 63–71.

19 In France in the eighteenth century, attempts were made by Buffon and Maupertuis among others to try and incorporate hereditary evidence, regeneration of lost organs and embryo development into universal theories of generation, but they could not formulate anything of sufficient force to overcome pre-existence. See Jacob, *Logic of living systems*, pp. 69–72.

20 G. Keynes (ed.), *Works of Sir Thomas Browne* (London, 1964), vol. 4, *Vulgar errors*, book 6, chapters 10 and 11.

21 Digby, *Two treatises*.

22 Digby, *Two treatises*, 1, pp. 213–27.

23 Nathaniel Highmore, *History of generation* . . . (London, 1651).
24 Highmore, *Generation*, pp. 32–3. Similar observations were often made by eighteenth-century savants in France, and Reaumur went so far as to describe experimental protocols for crossing poultry with different rump and claw characteristics. There is no evidence that such experiments were ever performed, however. Only Koelreuter seems to have done work of this sort before the late eighteenth century, but he confined his work to plants. See Jacob, *Logic of living systems*, p. 69; and R.C. Olby, *Origins of Mendelism* (Schocken Books, 1967), pp. 23–5.
25 Highmore, *Generation*, p. 32.
26 Jacob, *Logic of living systems*, pp. 76–7.
27 Highmore, *Generation*, p. 102.
28 In the preface to Francis Willoughby's *Ornithology*, Ray chastises his patron for describing the precise colour markings on the feathers of his specimens, since Ray doubted if any two individuals of the same species were ever identical for such characters. See J. Ray (ed.), *Ornithology of Francis Willoughby* (London, 1678), preface, p. 3.
29 J. Ray, *The wisdom of God manifested in the works of creation* (London, 1691), pp. 36–7, 217.
30 Ray, *Wisdom*, pp. 38–9.
31 Ray, *Wisdom*, p. 74.
32 Ray, *Wisdom*, pp. 217–19. In these passages Ray made reference to efficient and material causes, showing his Aristotelian attitude long after this was unfashionable.
33 Gasking, *Investigations*, p. 56.
34 Ray, *Ornithology of Francis Willoughby*, preface.
35 Derham, *Mr Ray's philosophical letters*, pp. 343–7.
36 One might suggest there is no *a priori* case for this. In a depressed market, anyone who by technical or managerial innovation can reduce the price of a product or service and still operate at a profit might be successful and indeed his actions and those of others like him be part of the framework for economic recovery. This would apply especially to the provision of basic commodities, such as foodstuffs and clothing, for which there is always a demand.
37 R.B. Outhwaite, *Inflation in Tudor and early Stuart England* (Macmillan, 1969), pp. 9–15.
38 Outhwaite, *Inflation*, pp. 40–1.
39 Outhwaite, *Inflation*, pp. 23–36.
40 Outhwaite, *Inflation*, pp. 37–47.
41 M.M. Postan, *Medieval economy and society* (Penguin, 1972), *passim*, and J.Z. Titow, *English rural society, 1200–1350* (Allen and Unwin, 1972), *passim*.
42 Data quoted in P. Ramsey, *Tudor economic problems* (Gollancz, 1963), pp. 15–16. Russell's data derive from his *British medieval population* (1948).
43 E.A. Wrigley and R.S. Schofield, *The population history of England, 1541–1871* (Edward Arnold, 1981); D.M. Palliser, 'Tawney's century 1540–1640', *EHR*, 2nd series, 35 (1982), 339–53.
44 Palliser's 'Tawney's century' may be taken as a statement of the current optimistic position.
45 Bridbury may perhaps be seen as the leading pessimist. See, for instance, A.R. Bridbury, *Economic growth: England in the later middle ages* (Harvester Press, 1975); and A.R. Bridbury, '16th-century farming', *EHR*, 2nd series, 27 (1974), 538–56.

46 M.W. Flinn, *Origins of the industrial revolution* (Longmans, 1966), chapter 2.
47 E.L. Jones (ed.), *Agriculture and economic growth in England, 1650–1815* (Methuen, 1967), introduction and chapter 6, both by Jones.
48 This is essentially the view adopted by Bridbury in '16th-century farming'.
49 E. Kerridge, *The agricultural revolution* (George Allen and Unwin, 1967) and C. Lane, 'The development of pastures and meadows during the 16th and 17th centuries', *AHR*, 28 (1980), 18–30.
50 J.H. Bettey, 'The development of water meadows in Dorset during the 17th century', *AHR*, 25 (1977), 37–43.
51 J. Thirsk, *The agrarian history of England and Wales* (Cambridge University Press, 1967), vol. 4, pp. 1–112; Kerridge, *Agricultural revolution*, pp. 41–180.
52 Master Fitzherbert, *The boke of husbandry 1533*, ed. W.W. Skeat (London, 1882), p. 42.
53 Thirsk, *Agrarian history*, vol. 4, pp. 161–99.
54 Thirsk, *Agrarian history*, vol. 4, p. 163. This prejudice was well expressed by John Aubrey in his diary in the 1640s. He contrasted the North Wiltshire dairy country with the South Wiltshire corn and sheep downs. He found the pastoral northerners dull-spirited, lazy and given to religious fanaticism. On the arable downs he found the peasantry fit, hard-working and untroubled by deep religious conviction. See Dick, *Aubrey's brief lives*, pp. 48–9.
55 Thirsk, *Agrarian history*, vol. 4, pp. xxx, xxxvi.
56 John Aylmer compared English and German diets as follows: 'They eat herbes: and thou Beefe and Mutton. Thei rotes: and thou butter, cheese and eggs. Thei drink commonly water: and thou good ale and beare. Thei go from market with a sallet: and thou with good fleshe fill thy wallet' (quoted by F.J. Fisher (ed.), *Essays in the economic and social history of Tudor and Stuart England* (Cambridge University Press, 1961), pp. 12–13).
57 L.A. Clarkson, 'The organisation of the English leather industry in the late 16th and early 17th centuries', *EHR*, 2nd series, 13 (1960), 245–56. Clarkson points out that up to one-fifth of the whole artisan labour force was involved in industrial production based on leather. Its significance is constantly underrated because there was no export trade to provide visual summaries of its activities. See L.A. Clarkson, *The pre-industrial economy in England, 1500–1750* (Batsford, 1971), pp. 81, 88–9.
58 Palliser, 'Tawney's century', *passim*.
59 W. Harrison, 'Description of England', book 3, chapter 1, in *Holinshed's chronicles of 1586* (London, 1807).

4 The horse: breeding for war, sport and fashion

1 Some time ago Peter Mathias suggested that agricultural livestock improvers may have learned from the Restoration enthusiasts for foxhounds and Arab horses. See Jones, *Agriculture and economic growth in England, 1650–1815* (Methuen, 1967), introduction, p. 11.
2 Thirsk, *Horses in early modern England*.
3 N. Morgan, *The horseman's honour or the beautie of horsemanship . . .* (London, 1620), p. 14.
4 G. Markham, *How to chuse, ride, train and diet both hunting and running horses . . .* (London, 1599), fols. A3 and A4. Blundeville made no mention of Arab horses in 1580. See Thirsk, *Horses in early modern England*, p. 20.
5 J. Ball, *The farmer's complete guide through all the articles of his profession* (London,

1760), pp. 372–3; T. Hale, *A compleat body of husbandry* . . . (London, 1756), pp. 199–200.

6 An exact parallel was outlined in Normandy, where it was considered more profitable to breed from coach than from saddle horses. See Anon., 'On the breeding of horses' in M. Peters (ed.) *De re rustica*, (1770), vol. 2, no. 7, article 60, pp. 90–1.

7 W. Ellis, *The modern husbandman* . . . (London, 1750), vol. 3, p. 173.

8 F. de la Rochefoucauld, *Mélanges sur l'Angleterre* (1784), trans. S.C. Roberts (ed. J. Marchant) as *A Frenchman in England* . . . (Cambridge, 1933), pp. 73–4; T. Pennant, *British zoology*, vol. 1 (London, 1768), p. 2.

9 See, for instance, J. Donaldson, *General view of the agriculture of the county of Northampton* (Edinburgh, 1794), pp. 52–3.

10 Ellis, *The modern husbandman*, vol. 3, p. 173; Hale, *Compleat body*, p. 201.

11 The designations 'trotting' and 'ambling' seemed to be general terms for describing cavalry and saddle horses respectively. See *APC*, vol. 6, 'Letter from the bailiff of Lowestoft concerning horses of possible illegal entry 2nd June 1556', pp. 37–8, and HMC, *MSS of the Earl of Ancaster* . . . (Dublin, 1907), pp. 453–6. Markham gave detailed instructions on the training of war horses, which he referred to as 'trotting' or 'charging' horses. See Markham, *How to chuse*.

12 D.M. Goodall, *A history of horse breeding* (Robert Hale, 1977), pp. 144–60.

13 Thus, the stable of the 5th Duke of Northumberland in 1512 contained two 'great double trotting horses', which were presumably war horses, and seven 'great trotting horses', which were described as used for the chariot, presumably a cart or waggon. It would seem that the same general type were therefore used as chargers and waggon horses. See R. Berenger, *The history and art of horsemanship* (London, 1771), vol. 1, p. 178.

14 Thirsk, *Horses in early modern England*, p. 10.

15 Thirsk, *Horses in early modern England*, pp. 7–11; I.S. Leadam, *Select cases in the star chamber: vol. 2, 1509–1544* (Seldon Society, 1911), p. 231.

16 *Statutes of the realm*, 22 Henry VIII c. 7; 23 Henry VIII c. 16; 27 Henry VIII c. 6; 31 Henry VIII c. 7; 33 Henry VIII c. 17.

17 For instance, *Statutes of the realm*, 1 Edward VI c. 5; 1 Elizabeth c. 18. See also the proclamation 'Against the carrying and conveying of horses and mares out of the realm' of 1562, in P.L. Hughes and J.F. Larkin (eds.), *Tudor royal proclamations* (Yale, 1969), vol. 2, pp. 200–1; and *APC*, vol. 10, pp. 131–2 (20 May 1579) and vol. 12, p. 339 (21 February 1580).

18 HMC, *Calendar of the manuscripts of the Marquis of Salisbury, preserved at Hatfield House* (London, 1883–1973), vol. 5, pp. 273, 414–15.

19 *Statutes of the realm*, 27 Henry VIII c. 6.

20 *Statutes of the realm*, 33 Henry VIII c. 5; Thirsk, *Horses in early modern England*, pp. 12–13.

21 Thirsk, *Horses in early modern England*, pp. 15–16. See also Hughes and Larkin, *Tudor royal proclamations*, vol. 2, 'For the having and keeping of horses and geldings, and for the furniture and having of armour and weapons' (7 May 1562), pp. 197–200. The organisation and administration of the musters became much more efficient during Elizabeth's reign, especially after 1580 as the threat of the Spanish invasion became more imminent. See, for instance, *CSPD* (1547–80), pp. 643–85.

22 *Statutes of the realm*, 32 Henry VIII c. 13.

23 *Statutes of the realm*, 8 Elizabeth c. 8.
24 C. M. Prior, *The royal studs of the sixteenth and seventeenth centuries . . .* (London, 1935), pp. 2, 11, 112.
25 Thirsk, *Horses in early modern England*, pp. 11–12; *CLPF & D*, Henry VIII, vol. 1, p. 930, vol. 2, pp. 63, 204, 278, 344–5, 409, 413, 935, vol. 15, *passim; APC*, vol. 6 (1552–4), pp. 212, 214.
26 The gentleman declined to join Henry's entourage and remained in the service of the marquis. See Prior, *The royal studs*, p. 1.
27 Thirsk, *Horses in early modern England*, pp. 16–17. Alexander de Bologna and Jacques de Granadoe appear to have been officers in the royal stables from 1526 to at least 1544. Sir Philip Sidney, the Earl of Leicester and Lord Walden also employed Italian horsemen. Jacques de Granadoe seems to have remained on the royal payroll until 1546. He possessed horses of his own listed separately from the royal animals in an inventory of that year. See *CLPF & D*, vol. 21, pp. 479–80.
28 The leading works of the Continental and largely Italian school were F. Grisone, *The rules for riding* (Naples, 1550); C. Corte, *The horse trainer . . .* (Venice, 1562); and M. Fugger, *How to manage a stud of good and noble war horses* (1578). The first English translation to appear was that of Thomas Blundeville of Grisone's work in about 1560. This was called a *Newe booke containing the arte of rydinge and breakinge greate horses* and was a severely modified version of the Italian text. The first edition of his greatly expanded book, *The fower chiefyst offices belonging to horsemanshippe*, containing discussions of breeding, appeared in 1565. Two other major translations of the Italian school came out in the sixteenth century: H. Bedingfield, *The art of riding* (1584), which was a translation of Corte; and J. Astley, *The art of riding* (1584), which was a translation of Grisone.
29 Arnold had imported Flemish horses in 1546, travelled in Italy in Edward VI's reign and thereafter maintained a stud of Neapolitan war horses. See Thirsk, *Horses in early modern England*, p. 17. Harrison noted that he had been a leading horse breeder and writer, while Blundeville said he had been a breeder of Neapolitan horses.
30 One of his forebears had been Henry Markham, Gentleman Pensioner to Henry VIII. His father was Robert Markham of Cotham in Nottinghamshire, a noted horse breeder and friend of the horse-fancying fraternity. See Thirsk, *Horses in early modern England*, p. 20.
31 Prospero d'Osma, 'Explanation and account of my sojourn at the stud at Malmesbury and then at Tutbury' (1576), in Prior, *The royal studs*, pp. 11–38, in translation.
32 Christopher Clifford, *The schoole of horsemanship* (London, 1585).
33 Harrison, 'Description of England', book 3, chapter 1.
34 Goodall, *Horse breeding*, chapter 6; Prior, *The royal studs*, pp. 86–7; William Cavendish (Duke of Newcastle), *A new method and extraordinary invention to dress horses . . .* (London, 1667), p. 30.
35 Blundeville, *Fower chiefyst offices*, book 1, chapter 2.
36 Markham justified the system of hybridisation to produce working horses by arguing that the worst features of any breed were eliminated, a somewhat naive assumption. See G. Markham, *Cavalarice or the English horseman* (London, 1625), p. 18.
37 Blundeville complained that hybridisation had made it difficult to say whether particular horses were of pure exotic stock or not. D'Osma said the pure breeds

were natural and possessed specific qualities which were lost by cross-breeding. Such cross-breds were 'bastard' horses. See Blundeville, *Fower chiefyst offices*, fol. 5; and d'Osma, 'Explanation and account', p. 75.

38 Markham, *Cavalarice*, p. 18.

39 Blundeville, *Fower chiefyst offices*, book 1, chapter 1; d'Osma, 'Explanation and account', pp. 49–53; Markham, *How to chuse*, fols. A1 and A2. The belief that soil and climatic conditions directly influence the hereditary constitution of a horse still persists among modern commentators and historians. See A.A. Dent, *Cleveland bay horses* (J.A. Allen, 1978), pp. 28–30.

40 But Markham did not accept that his objection had been overcome. He said the constant stimulus of a mare every three days would overheat the stallion more than running him with a group of mares and so do him even more harm. See Markham, *Cavalarice*, chapter 5.

41 Indeed, he disapproved strongly of such practice even if a groom-assisted mating was necessary. See Markham, *Cavalarice*, p. 29.

42 'Still' here meaning in relation to classical practice, where he had read of the same 'unnatural' phenomena.

43 Blundeville, *Fower chiefyst offices*, book 1, chapters 5 and 6; Clifford, *Schoole of horsemanship*, unpaginated; Markham, *How to chuse*, fols. A2 and A3, and *Cavalarice*, chapter 5; d'Osma, 'Explanation and account', pp. 100–3.

44 Markham, *How to chuse*. However, Markham could not bring himself to continue with this advice. In *Cavalarice* he sensibly remarked that Blundeville's system implied breeding from the best rarely, and from the worst often, which was obviously unsound. By this time Markham was beginning to find some of the more unrealistic advice of the Italians irksome. See Markham, *Cavalarice*, chapter 9.

45 This general emphasis on keeping the female lean at mating was found for all species of animal. D'Osma explicitly stated that brood mares should at no stage be overfed. See d'Osma, 'Explanation and account', pp. 58–60, 94–9.

46 Blundeville, *Fower chiefyst offices*, book 1, chapter 15; d'Osma, 'Explanation and account', p. 61.

47 Clifford, *Schoole of horsemanship*: 'I have also ridden on an horse of the Marshal Byrone's begotten of a Turkish stallion that was 24 years of age, and I never found in all my life a better horse for travel and service.'

48 Clifford, *Schoole of horsemanship*; Blundeville, *Fower chiefyst offices*, book 1, chapter 4; Markham, *Cavalarice*, pp. 26–7; d'Osma, 'Explanation and account', pp. 77–86.

49 By some the phase of the moon was said to govern the sex of offspring in horse breeding. Several quasi-scientific experiments were reported in the literature of attempts to confirm or deny this phenomenon. See Fitzherbert, *Boke of husbandry*, pp. 57–61; Blundeville, *Fower chiefyst offices*, chapter 7. Markham was sceptical about astrological and imaginative influences over the determination of sex in the offspring. See Markham, *Cavalarice*, chapter 8.

50 Blundeville, *Fower chiefyst offices*, book 1, fol. 13a.

51 Blundeville, *Fower chiefyst offices*, book 1, fol. 33.

52 Markham, *Cavalarice*, p. 26.

53 Thus Blundeville spoke at length about the desirable qualities and points of stallions, but hardly mentioned the desirable points of mares. Markham laid much more stress on brood mares. See Markham, *How to chuse*, fol. A3. He thought disposition and fertility to be their most important features and

relegated their shape and size to secondary status. See Markham, *Cavalarice*, pp. 28–9.

54 Blundeville admitted that all his views on horse aesthetics were entirely derived from Grisone. See *Fower chiefyst offices*, book 1, fol. 13a.

55 Blundeville, *Fower chiefyst offices*, book 2, fols. 1 and 1a.

56 Blundeville, *Fower chiefyst offices*, book 2, fols. 2a and 3.

57 Blundeville, *Fower chiefyst offices*, book 1, fol. 14.

58 Markham, *Cavalarice*, pp. 21–2.

59 Pedigree breeding in hounds seems to be older. See G. Turbeville, *The noble arte of venerie* (1576) (Oxford, 1908), pp. 4–6.

60 Thirsk, *Horses in early modern England*, pp. 13–14; *CLPF & D*, Henry VIII, vol. 2 (1519), p. 63.

61 These were the only studs surveyed by Prospero d'Osma in 1576.

62 *CLPF & D*, Henry VIII, vol. 21 (1546), pp. 479–80.

63 HMC, *MSS of the Earl of Ancaster*, pp. 453–6.

64 Naming horses in this way was common in the sixteenth century. Many animals were designated by their original owner in the stable of Sir Francis Walsingham. See Thirsk, *Horses in early modern England*, p. 16.

65 Thirsk, *Horses in early modern England*, p. 9.

66 Thirsk draws attention to the origin of records about individual horse ancestry in other stables. See *Horses in early modern England*, p. 16.

67 D'Osma, 'Explanation and account', pp. 56–7.

68 D'Osma, 'Explanation and account', pp. 85–6.

69 Prior, *The royal studs*, pp. 10, 56.

70 D'Osma described the available parkland as a five-mile circumference park at Castlehay, divided into ten regions, a two-and-a-half-mile park at Stockley, a two-mile park at Rolleston, two additional small parks and two small meadows for hay. In 1619 the stud grounds consisted of Castlehay of 990 acres, Hanbury Park of 373 acres, the little park at Tutbury Castle of 71 acres and the two small closes. See Prior, *The royal studs*, pp. 21–4, 45–6. However, the park was not exclusively given over to horse breeding. In 1598 all sorts of cattle other than horses were recorded there. See HMC, *Salisbury MSS*, vol. 9, pp. 114–15.

71 HMC *Salisbury MSS*, vol. 9, pp. 114–15.

72 *CSPD*, Elizabeth, vol. 240 (September–December 1591), p. 133.

73 HMC, *Salisbury MSS*, vol. 8, p. 391.

74 Conrad Gesner, *Historia animalium* (5 vols., 1551–87).

75 Edward Topsell, *The historie of four-footed beasts and serpents . . .* (London, 1607).

76 Prior, *The royal studs*, chapter 1.

77 Prior, *The royal studs*, chapter 1.

78 For instance, Robert Reyce, describing Suffolk horses in the early seventeenth century, was firmly of the opinion that type was governed by environment. The rich lowland soils bred a sluggish horse, fit only for draught and burden while the upland regions bred cavalry horses. See Lord Francis Hervey (ed.), *The breviary of Suffolk by Robert Reyce of 1618* (London, 1902), pp. 42–4.

79 Morgan, *The horseman's honour*, p. 2.

80 Morgan, *The horseman's honour*, p. 20.

81 Morgan, *The horseman's honour*, pp. 20–1.

82 Morgan, *The horseman's honour*, p. 22.

83 Morgan, *The horseman's honour*, pp. 23–8.

84 Morgan, *The horseman's honour*, pp. 31–2.

85 Morgan, *The horseman's honour*, p. 42.
86 Morgan, *The horseman's honour*, pp. 44–53.
87 Morgan, *The horseman's honour*, pp. 57–63.
88 Morgan, *The horseman's honour*, pp. 63–86.
89 Morgan, *The horseman's honour*, pp. 97–101.
90 Morgan, *The horseman's honour*, pp. 104–11.
91 Morgan, *The horseman's honour*, p. 117.
92 Specifically, however, Newcastle said he had no time for the view that the Earth and its creatures were constantly degenerating. Thus, 'some people would make us believe that the Barbs and Spanish Jenets beget too small colts, in respect that nature daily dries up and becomes elder . . . As to what people say that nature dries daily up, I believe the sun is as hot at present, as it was the first moment it was created, and that the Earth is also as fertile.' See Sir W. Hope, *The perfect mareschal or compleat farrier . . . of sieur de Solleysel* (Edinburgh, 1746).
93 Hope, *The perfect mareschal*.
94 Cavendish, *A new method*, pp. 23–4.
95 Cavendish, *A new method*, p. 26.
96 Cavendish, *A new method*, pp. 28–9.
97 Cavendish, *A new method*, p. 88; Hope, *The perfect mareschal*, pp. 206–7.
98 Hope, *The perfect mareschal*, p. 205.
99 Cavendish, *A new method*, pp. 87–8. Nor should the stallion be too old 'except necessity force you'.
100 Cavendish, *A new method*, pp. 87–92.
101 Cavendish, *A new method*, p. 93.
102 Hope, *The perfect mareschal*, p. 218.
103 Hope, *The perfect mareschal*, p. 220.
104 Apart from Markham's advocacy of Arab stallions in 1607, Michael Barret, in 1618, had recommended both Barb and Turkish stallions as the best kinds of horse for producing general purpose animals. See Prior, *The royal studs*, p. 107, quoting *An hipponomie or the vineyard of horsemanship*.
105 Cavendish, *A new method*, pp. 50–1, and 56–7; Hope, *The perfect mareschal*, p. 211.
106 Cavendish, *A new method*, p. 58.
107 Cavendish, *A new method*, pp. 64–6.
108 Cavendish, *A new method*, pp. 67–9.
109 Cavendish, *A new method*, pp. 107–8; Hope, *The perfect mareschal*, p. 211.
110 Prior, *The royal studs*, p. 104.
111 Prior, *The royal studs*, pp. 39–41. Nothing remains of Villiers's own stud records at Burley-on-the-Hill near Oakham.
112 Prior, *The royal studs*, p. 42.
113 Prior, *The royal studs*, pp. 66–8, 48–9.
114 For details, see Russell, 'Animal breeding', pp. 189–90, 377.
115 For details, see Russell, 'Animal breeding', pp. 377–8.
116 Scott-Robinson, 'The expense book of James Master'; Prior, *The royal studs*, p. 55.

5 Horse breeding in the eighteenth century: blood, speed and carriages

1 Markham, *How to chuse*, chapter 3.
2 Markham, *How to chuse*, chapter 4. The sport became fashionable among the gentry from 1580 onwards. See J. Thirk, *Horses in early modern England*, p. 22.

3 Markham, *How to chuse,* fol. J1.
4 HMC, *12th Report appendix part IV,* 'Manuscripts of his grace the Duke of Rutland, preserved at Belvoir Castle' (London, 1888–1905), vol. 4, pp. 409–10, 420.
5 HMC, *Rutland MSS,* vol. 4, pp. 442–3, 454–5, 502–3, 513.
6 HMC, *Salisbury MSS,* vol. 19, p. 283; HMC, *Rutland MSS,* vol. 1, p. 454.
7 HMC, *Salisbury MSS,* vol. 22, p. 334.
8 J. Britten (ed.), *John Aubrey's natural history of Wiltshire* (Wiltshire Topographical Society, London, 1847), pp. 87–8.
9 HMC, *Rutland MSS,* vol. 2, p. 337 and vol. 4, p. 545.
10 HMC, *14th Report appendix part VII,* 'Manuscripts of the city of Lincoln' (London, 1895), p. 106; HMC, *Rutland MSS,* vol. 2, p. 128.
11 HMC, *Rutland MSS,* vol. 4, pp. 545–6, 550–1.
12 HMC, *Rutland MSS,* vol. 4, pp. 556–7.
13 S.H.A. Hervey (ed.), *The diary of John Hervey, 1st Earl of Bristol, with extracts from his book of expenses, 1688–1742* (Wells, 1894).
14 Prior, *Racing calendar,* p. 76.
15 Prior, *Racing calendar;* P. Willett, *An introduction to the thoroughbred* (Stanley Park, 1975), pp. 128–30.
16 Cavendish, *A new method,* pp. 62–3.
17 Willett, *Introduction to the thoroughbred,* pp. 31–3; Prior, *Racing calendar,* pp. 128–9, 133, 135, 150, 152–3; Hervey, *Diary of John Hervey,* pp. 14, 27–8, 56; La Rochefoucauld, *Mélanges,* p. 65.
18 Arab blood reached the zenith of its success in about 1750, but its influence seems to have waned rapidly. Most commentators, noticing the fall in demand for Oriental stallions, have supposed that this correlated with the achievement of excellence by the native Thoroughbred, which imports could no longer improve. See Prior, *Racing calendar,* pp. 122–3. Others have been unable to see why Oriental sires were seldom themselves successful and therefore precisely what stimulus caused their continued import. See Willett, *Introduction to the thoroughbred,* pp. 35–6.
19 The famous early eighteenth-century racehorse, *Flying Childers,* was in a class of his own in trial matches against his contemporaries. Two independent time trials claimed that over distances of about four miles he achieved average speeds of 33 and 34 miles per hour. Even if these times are held to be optimistic, there is no ground for supposing, as many turf historians do, that the watches of the judges were grossly inaccurate. Over $2\frac{3}{4}$ miles in the Brown Jack Stakes on 18 July 1958, *Birthday Present* set up a record at approximately the same speed. The general run of modern Thoroughbreds are probably closer to *Birthday Present* than early eighteenth-century racers were to *Flying Childers.* See Willett, *Introduction to the thoroughbred,* pp. 24–5; Prior, *Racing calendar,* p. 130; and Pennant, *British zoology,* p. 3.
20 La Rochefoucauld, *Mélanges,* p. 63; W. Osmer, *A treatise on the diseases and lameness of horses . . .* (London, 1761), p. 247.
21 Prior, *The royal studs,* pp. 88–95; Cavendish, *A new method,* pp. 52–8, 71–2; HMC, *13th Report appendix,* 'Manuscripts of the Duke of Portland at Welbeck' (London, 1891–), vol. 2, pp. 179, 155–6, 259–60, 262; HMC, *11th Report appendix part IV,* 'Manuscripts of the Marquess of Townsend' (London, 1887), pp. 392–3.
22 Prior, *Racing calendar,* p. 12.
23 The use of a detailed pedigree system to record this purity of lineage certainly

owed a great deal to the practice of Arabian breeders, who kept lengthy genealogies of their high-bred animals with the objective of maintaining the purity of the breed. See G.L. Buffon, *The natural history of the horse . . .* (London, 1762), pp. 70–2; and Berenger, *History and art of horsemanship*, vol. 1, pp. 116–18. In the seventeenth century Newcastle had heard legendary tales of these horses and the Arab practice of keeping strict genealogies. His astonishment shows that even such a leading horseman as he had not yet found the maintenance of such records worthwhile. See Cavendish, *A new method*, pp. 72–3.

24 However, 'blood' in this context may have undergone a change of meaning. Discussion of 'blood' horses and pedigrees from the mid-eighteenth century shows quite clearly that the term was equated with some form of hereditary transfer. Originally, in the seventeenth century, 'blood' may merely have implied vigorous, vital, courageous, hot animals from warmer climates.

25 In the northern counties especially, small races were often run by Galloways and this practice persisted well into the early part of the eighteenth century. See J.J. Bagley and F. Tyrer (eds.), 'Blundell's diary, the great diurnall of Nicholas Blundell of Crosby, 1702–1728', *Record Society of Lancashire and Cheshire*, 110, 112, 114 (1968, 1970, 1972), 112 and 114, *passim*.

26 Hervey, *Diary of John Hervey*.

27 This mare was probably so called because Somerset's family name was Gordon. It is also possible that it was a Scottish Galloway.

28 WSRO, Misc. papers 1441. Transcribed by W.E. Thurgood as 'May it please your grace', PHA 6323–5, no. 3, no. 5, no. 6a; PHA 6322, no. 1; PHA 6323–5, no. 102. In his discussions of the correct form of a racehorse, Osmer mentioned *Cartouche* as one with a particularly good shape. He added that the animal was a Galloway. Nevertheless, Lord Godolphin had two mares sired by *Cartouche* in his stud. In addition, three of the mares which had originally belonged to Edward Coke before they were acquired by Godolphin, including the famous mare, *Roxana*, had been sired by an animal called the *Bald Galloway*, who was himself sired by the *St Victor Barb*, presumably off a Galloway mare. See Osmer, *Treatise*, p. 221; 'Stud books of Edward Coke and Lord Godolphin', in Prior, *The royal studs*, pp. 129–79.

29 Hervey, *Diary of John Hervey*.

30 The three foundation sires were *Matchem*, foaled in 1748, *Herod*, foaled in 1758 and *Eclipse*, foaled in 1764. See Willett, *Introduction to the thoroughbred*, p. 32.

31 Prior, *The royal studs*, p. 95.

32 Willett, *Introduction to the thoroughbred*, pp. 83–7. This may have been the origin of Marshall's opinion that inbreeding had already been tried out at Newmarket. See W. Marshall, *The rural economy of the Midland counties* (London, 1790), vol. 1, p. 300. Bakewell himself did not believe that inbreeding on any scale had ever been part of the technique of Thoroughbred breeders. See A. Young, 'A ten days tour to Mr Bakewell's', *Annals of Agriculture*, 6 (1786), 488.

33 The mare concerned was *Crofts Partner Mare*, by *Makeless* out of *Brown Farewell*. *Makeless* was *Brown Farewell's* sire, so this was a sire–daughter mating. See 'Lord Godolphin's stud book', in Prior, *The royal studs*.

34 Both pedigrees may be found in Willett, *Introduction to the thoroughbred*, pp. 25–6, 162. On the *General Stud Book* pedigree the dam of *Flying Childers*, *Bettey Leedes*, had an inbreeding coefficient of 0.38.

35 Of those kept between 1732 and 1754, only 6 from a total of 27 mares seem to

have been home-bred. See 'Lord Godolphin's stud book', in Prior, *The royal studs.*

36 D. Defoe, *A tour through the whole island of Great Britain, 1724–1726* (London, 1971), p. 400.

37 Defoe, *Tour,* p. 512.

38 R. Bradley, *A complete body of husbandry* (London, 1727), pp. 345–6, 348–9.

39 Hale, *Compleat body,* p. 202. Rochefoucauld noted that foals ran loose until they were three years old, but from that time they were constantly stabled, carefully dieted and exercised, each one being allocated a groom. See La Rochefoucauld, *Mélanges,* p. 64.

40 R. Wall, *A dissertation on the breeding of horses upon philosophical and experimental principles* . . . (London, 1758), pp. 61–2.

41 Wall, *Dissertation,* pp. 19–22.

42 Wall, *Dissertation,* pp. 22–7.

43 Wall, *Dissertation,* pp. 30–5.

44 Wall, *Dissertation,* pp. 39–41.

45 Wall, *Dissertation,* pp. 42–54, 63–4.

46 Wall, *Dissertation,* pp. 86–90.

47 Osmer, *Treatise,* pp. 268–73.

48 Osmer, *Treatise,* pp. 199–228, *passim.* He was led to believe in the extreme lightness of form and limb, 'cat-leggedness' as some of the sportsmen called it, by his view that since muscle and sinew seemed to be contiguous, they were composed of the same material, in other words that tendons could substitute for muscle. See Osmer, *Treatise,* pp. 221–4.

49 Osmer, *Treatise,* pp. 229–31.

50 Osmer, *Treatise,* pp. 234–5.

51 Osmer, *Treatise,* pp. 247–50.

52 Osmer, *Treatise,* p. 253.

53 J. Worledge, *Systemma agriculturae* (1675), and Harleian Miscellany VIII (1746), 'The grand concern of England explained' (1673), both in J. Thirsk and J.P. Cooper (eds.), *17th-century economic documents* (Oxford, 1972), pp. 165, 371–84; C. Merret, 'An account of several observables in Lincolnshire . . .', *Phil. Trans. Royal Society,* 19 (1695–7), 349; Britten, *Aubrey,* p. 61.

54 J. Mortimer, *The whole art of husbandry* . . . (London, 1707), pp. 149–50.

55 J. Laurence, *A new system of agriculture* (London, 1726), p. 129; Defoe, *Tour,* pp. 408–9.

56 Bradley, *Complete body of husbandry,* p. 356; R. Bradley, *The gentleman and farmer's guide for the increase and improvement of cattle* (London, 1729), pp. 265–6.

57 Marshall, *Midland counties,* vol. 1, pp. 306–9; R. Trow-Smith, *A history of British livestock husbandry* (Routledge, 1957 and 1959), vol. 2, pp. 159–60.

58 Marshall, *Midland counties,* vol. 1, pp. 311–12, 306–7, and vol. 2, pp. 84–5; T. Brown, *General view of the agriculture of the county of Derbyshire* (London, 1794), p. 25. Oxfordshire was still involved in the transfer trade of Blacks to London in the 1790s, while Shropshire to the west was largely supplied from Derbyshire and Leicestershire in the same period. See R. Davis, *General view of the agriculture of the county of Oxfordshire* (London, 1794), p. 25; and J. Bishton, *General view of the agriculture of the county of Shropshire* (London, 1794), pp. 10–11.

59 A. Young, *The farmer's tour through the east of England* (London, 1771), vol. 1, pp. 56–7; Donaldson, *Agriculture of Northampton,* pp. 51–2.

60 T. Stone, *General view of the agriculture of the county of Lincoln* (London, 1794), p. 62. General eighteenth-century prices for coach horses may be obtained from contemporary farm accounts. They were available at between £12 and £25. By contrast, ordinary cart horses were worth only £4 to £6. See HFRC Reading, Han 10/1/1 and Sal 5/1/1; Berks RO, D/EPb E7; Bucks RO, Drake MSS D/Dr/2/61; KAO, U593 Tylden MSS A3 and U23 Culpepper MSS E7; and HMC *Townsend MSS*, p. 202.

61 Stone, *Agriculture of Lincoln*, p. 62.

62 Certainly, just to the north in Durham, saddle horses for chapmen were the main horse export in 1704. See HMC, *Portland MSS*, vol. 2, p. 188.

63 *Jalap* was out of the mare, *Red Rose*, by the very successful racehorse, *Regulus*. See Dent, *Cleveland bay horses*, p. 51.

64 W. Marshall, *The rural economy of Yorkshire* (London, 1788), vol. 2, pp. 162–4, 172–3.

65 J. Tuke, *General view of the agriculture of the North Riding of Yorkshire* (London, 1794), p. 66. The same pursuit of fashion had also affected Northamptonshire in the last quarter of the century, when the county suffered from an influx of Thoroughbred stallions. See Donaldson, *Agriculture of Northampton*, pp. 52–3.

66 The change in fashion was undoubtedly associated with the expansion of the turnpike trusts during the eighteenth century, which resulted in general road surfaces being sufficiently improved for carriages to be handled by teams of light horses.

67 W. Marshall, *The rural economy of Norfolk* (London, 1787), vol. 1, p. 43; Young, *East of England*, vol. 2, pp. 173–5; A. Young, *General view of the agriculture of Suffolk* (London, 1794), pp. 42–3; La Rochefoucauld, *Mélanges*, p. 75.

68 Marshall, *Norfolk*, vol. 1, pp. 44–8.

69 J. Bailey and G. Culley, *General view of the agriculture of the county of Northumberland* (London, 1794), p. 15.

70 La Rochefoucauld, *Mélanges*, p. 74.

71 Defoe, *Tour*, p. 255; W. Marshall, *The rural economy of Gloucestershire* (Gloucester, 1789), vol. 1, p. 207. Dorset farmers were not particularly interested in their rather indifferent cart teams. See J. Claridge, *General view of the agriculture of the county of Dorset* (London, 1793), p. 12. Worcestershire seemed to breed its own type of cart and plough animal. The county did not avail itself of supplies from the specialist Midland breeders, so their grounds were too heavily stocked with brood mares. See W. Thomas, *General view of the agriculture of the county of Worcestershire* (London, 1794), pp. 11–12.

72 Ball, *Farmer's complete guide*, p. 374.

73 Bailey and Culley, *Agriculture of Northumberland*, p. 15.

74 Stone, *Agriculture of Lincoln*, pp. 62–3; Marshall, *Norfolk*, p. 44; Goodall, *Horse breeding*, pp. 214–16.

75 Ellis, *The modern husbandman*, vol. 3, p. 173. Marshall also found it difficult to exclude the influence of climate and nutrition from the excellence of Yorkshire saddle horses because he claimed that attempts to transfer breedstock into Norfolk had repeatedly failed and he could detect no special merit in the methods then being used to breed horses in Yorkshire. See Marshall, *Yorkshire*, vol. 2, pp. 160–1, 166–7.

76 Mortimer, *Husbandry*, pp. 149–50; Hale, *Compleat body*, pp. 151–2; G. Jacob, *The country gentleman's vade mecum . . .* (London, 1717), p. 12.

77 Mortimer, *Husbandry*, pp. 151–2; Jacob, *Vade mecum*, p. 12; Bradley, *Complete*

body of husbandry, p. 344; Bradley, *Gentleman and farmer's guide*, pp. 280–1; Ellis, *The modern husbandman*, vol. 3, pp. 173–4.

78 Marshall, *Yorkshire*, vol. 2, pp. 163–4, 166–7; Tuke, *Agriculture of Yorkshire*, p. 67.

79 Ruricolo Glocestris, 'A letter to the editors on the necessity of breeding colts from sound stallions', *Museum Rusticum*, 2: 79 (1764).

80 Laurence, *System of agriculture*, pp. 129–30.

81 Bradley, *Gentleman and farmer's guide*, p. 281.

82 Ellis, *The modern husbandman*, vol. 3, p. 174.

83 Anon., 'On the pernicious practice of breeding from blind stallions', *Museum Rusticum*, 3: 41 (1765).

84 Berks RO., D/EHr E2.

85 Yorkshire farm breeders of the 1780s, despite their long reputation for horse breeding, spent no time or effort in assembling suitable mares, while taking good care to use only quality stallions. See Marshall, *Yorkshire*, vol. 2, pp. 166–7.

86 Marshall, *Yorkshire*, vol. 2, pp. 172–3; Marshall, *Midland counties*, vol. 2, pp. 83–4.

87 Marshall, *Midland counties*, vol. 1, pp. 303–5.

88 For seventeenth-century examples, see Russell, 'Animal breeding', pp. 200–1.

89 Anon., 'The will of Nicholas Blundell 1736', *Trans. Hist. Soc. of Lancs. and Cheshire*, new series, 30 (1915), 261–4.

90 Bagley and Tyrer, 'Blundell's diary', 112 and 114, *passim*.

91 KAO, U1127 Smith Masters MSS A7.

92 KAO, U145 Faunce MSS A4/1.

93 Advertisement in the *Kentish Post*, 23 April 1746. See D.A. Baker, 'Agricultural prices, production and marketing with special reference to the hop industry in north-east Kent, 1680–1760', unpublished Ph.D. thesis, University of Kent (1976), p. 351.

94 HFRC, Reading Sal 5/1/1.

95 KAO, U471 Waldershare Park MSS A29.

96 KAO, U301 Misc. deeds E6. Yorkshire stallions travelling in Kent were also recorded in the 1780s, covering at the charge of 15s and 2s 6d for the groom. See KAO, U769 Snell MSS E1.

97 G.D. Lumb (ed.), 'Extracts from the *Leeds Intelligencer* and the *Leeds Mercury* 1769–1776', *Thoresby Society*, 33 (1938), 94. Thoroughbred stallions began to be advertised for stud work in the Racing Calendars from 1744. Average Thoroughbreds were at 1 guinea and by the mid-1750s, 3 guineas was standard for the most fashionable sires. See Prior, *Racing calendar*, pp. 121–2. By the 1770s and 1780s the most noted racers, such as *Eclipse* and *Highflyer*, were commanding fees of between 15 and 20 guineas a mare. See Willett, *Introduction to the thoroughbred*, pp. 32–3. The general run of Thoroughbred stallions available for ordinary farmers and gentlemen in the 1780s were second-rank animals. Winners of classic races were preserved exclusively for covering Thoroughbred mares for some two or three years before being downgraded to travelling or covering at second-rate prices. This was done specifically to offset some of the huge costs of racing. See La Rochefoucauld, *Mélanges*, pp. 65–6.

98 HFRC Reading, microfilm, P.262.

99 Berks RO, D/EBu A10.

100 H. Home (Lord Kames), *The gentleman farmer* . . . (London, 1776), p. 39; Philip, 'On the choice of horses and mares for breeding', *The Scots Farmer*, 1 (1773), 321–2.
101 KAO, U769 Snell MSS E1; ESRO, Dunn MSS 38/7.
102 Young, *East of England*, vol. 1, p. 57. In the 1790s Midland Black stallions travelling from their home region for the season were being used by cart horse breeders in the Isle of Sheppey. See J. Boys, *General view of the agriculture of the county of Kent* (London, 1794), p. 70.
103 Young, *East of England*, vol. 1, pp. 119–20.
104 Donaldson, *Agriculture of Northampton*, p. 53.

6 Cattle breeding: dairymen, graziers and the techniques of their 'fancy'

1 G.H. Tupling, *The economic history of Rossendale* (Manchester University Press, 1927); H.J. Hewitt, 'Medieval Cheshire', *Chetham Society Publications*, new series, 88 (1929); R.A. Donkin, 'Cattle on the estates of medieval Cistercian monasteries in England and Wales', *EHR*, 2nd series, 15 (1962), 31–53.
2 A.R.B. Haldane, *The drove roads of Scotland* (David and Charles, 1973), pp. 180–3; E. Melling (ed.), *Kentish sources* (Maidstone, 1961), vol. 3, pp. 13–14; A.C. Lodge (ed.), *The account book of a Kentish estate, 1616–1704* (London, 1927); J. Cornwall, 'Farming in Sussex, 1560–1640', *Sussex Arch. Collections*, 92 (1954), 76–7; Trow-Smith, *Livestock husbandry*, vol. 1, p. 191.
3 Thirsk, *Agrarian history*, vol. 4, pp. 83–9; Hervey, *Breviary of Suffolk*, p. 40; J. Houghton (ed.), *A collection for the improvement of husbandry and trade* (London 1692–1703), 7: 149 (7 June 1695) and 7: 154 (12 July 1695); Britten, *Aubrey*.
4 The traditional stock-fatting regions were supplemented during the seventeenth century by the transition of Midland grazing regions from sheep walk to cattle finishing. See J. Thirsk, 'Agrarian history, 1540–1950' in W.G. Hoskins and R.A. McKinley (eds.), *Victoria history of the county of Leicestershire* (Oxford University Press, 1954), pp. 220–1. For the rising importance of permanent pasture for animal grazing in the Midlands after 1650, see J. Broad, 'Alternate husbandry and permanent pasture in the Midlands, 1650–1800', *AHR*, 28 (1980), 77–89.
5 Restrictions were eased on the export of livestock of all types (apart from sheep where the clothing interests opposed it) in the seventeenth century. By 1663 the grazing of stores for fat beef had become a trade with a powerful lobby, since the import of Irish, ready-fatted cattle was banned in that year and the ban reiterated in later years. See *Statutes of the realm*, 15 Charles II c. 7; 18 Charles II c. 2; 20 Charles II c. 7; 22 Charles II c. 13; 32 Charles II c. 2; and C.H. Firth and R.S. Rait (eds.), *Acts and ordinances of the interregnum* (London, 1911), vol. 1, pp. 1043–8.
6 Fitzherbert, *Boke of husbandry*, pp. 52–3.
7 Harrison, 'Description of England', book 3, chapter 1, p. 370.
8 G. Markham, 'The English housewife' (1637), chapter 6 in *A way to get wealth* (London, 1638); T. Tusser, *Five hundred points of good husbandry* (London, 1610), January's husbandry, verse 39.
9 G. Markham, 'Cheap and good husbandry' (1631), pp. 688–707 in *A way to get wealth*; Markham, 'English housewife', pp. 190–206.
10 Harrison wrote of the fancy of horn form in the Longhorn, noting attempts to mould the shape by annointing young horns with honey. See Harrison,

'Description of England', p. 370; and W. Smith, *The vale royal of England* (1656) quoted in G.E. Fussell, 'Four centuries of Cheshire farming systems, 1500–1900', *Trans. Hist. Soc. Lancs. and Cheshire*, 106 (1954), 60. By 1694 Houghton recorded the art of altering horn shape had graduated to moulding them with hot irons while they grew. See Houghton, *Collection*, 5: 115 (12 October 1694).

11 Trow-Smith, *Livestock husbandry*, vol. 1, pp. 202–5.

12 Markham's three major cattle types implied three distinct colour patterns, but most regions in the sixteenth and seventeenth centuries evidently had very mixed herds. In the North Wiltshire dairy zone in the seventeenth century, pied cattle were the commonest but black, brown and red cattle were also frequent. Cattle bought in Ross, Bridgenorth, Worcester and Stratford markets in 1626–7 were predominantly black in colour, but there were also plenty of brown, red, pied, yellow and white animals. A similar pattern of very mixed colour was also normal in the sixteenth-century herds in the northern counties, judged from the collections of printed inventories. See Britten, *Aubrey*, p. 61; and KAO, U269 Sackville MSS A413.

13 This data has been comprehensively reviewed by Trow-Smith, *Livestock husbandry*, vol. 2, pp. 84–120. The following account relies heavily upon this review.

14 G. Culley, *Observations on livestock* (4th edn, London, 1807), pp. 23–5.

15 However, in the early 1700s Lincolnshire was still the most famous Shorthorn region. See Defoe, *Tour*, p. 413.

16 Mortimer, *Husbandry*, p. 166.

17 J. Monk, *General view of the agriculture of the county of Leicestershire* (London, 1794), pp. 34–6; W. Pitt, *General view of the agriculture of the county of Leicestershire* (London, 1809), pp. 217–18; HFRC Reading, Lei 4/1/2.

18 Hale, *Compleat body*, p. 209.

19 Berks RO, D/ED E48 C. This estimate is based on the 49 mature animals in the Michaelmas 1761 stock list. Multiple colour patterns in herds were recorded in virtually every case where they were stated. For instance, KAO, U1127 Smith Masters MSS A7; KAO, U269 Sackville MSS A45/2; ESRO, Hooke MSS 16/3; LAO, Dixon MSS 4/1; KAO, U593 Tylden MSS A3.

20 Ellis, *The modern husbandman*, vol. 4, p. 143.

21 Marshall, *Yorkshire*, pp. 181–4.

22 Defoe, *Tour*, p. 513.

23 Laurence, *System of agriculture*, p. 130. Culley also remembered a gentleman of Durham (Mr Dobison) who had initiated the import of Dutch Shorthorns in his youth (before 1750). See Culley, *Observations*, pp. 29–30.

24 H. Berry, *Improved shorthorns and their pretensions stated, being an account of this celebrated breed of cattle* (2nd edn, London, 1830), pp. 13–15; Anon. 'Housing of great cattle', *Museum Rusticum*, 5: 25 (1765), 126.

25 Hale, *Compleat body*, p. 211.

26 Anon. 'An account of the extraordinary size of neat cattle and sheep in Teesvale and the neighbouring Yarm', *Museum Rusticum*, 6: 47 (1766) and 6: 31 (1766). The Staffordshire bull carcass weight was 1665 lb. The forequarters were grossly heavier than the hind and the hide was very heavy at 172 lb, suggesting a Longhorn.

27 Many examples for the late eighteenth and early nineteenth centuries were quoted by Culley and Berry. See Culley, *Observations*, pp. 33–4; and Berry, *Improved shorthorns*, pp. 20–9.

28 Ellis, *The modern husbandman*, vol. 4, chapter 14.
29 Hale, *Compleat body*, pp. 209–16; Ellis, *The modern husbandman*, vol. 4, pp. 143–8; 'Country Gentleman', *The complete grazier: or gentleman and farmer's directory* (London, 1776), pp. 4–6; Pennant, *British zoology*, vol. 1, pp. 17–18.
30 Harrison, 'Description of England', pp. 186, 370.
31 This is based on the conventional modern estimate that dead weight = 60% of live weight. Trow-Smith, however, assumed that 672 lb was a live weight value. See *Livestock husbandry*, vol. 1, p. 177. Thorold Rogers published some actual purchase figures for naval procurement in London between 1546 and 1548 which showed somewhat smaller animals, between 307 and 448 lb dead weight, corresponding to live weights of about 500 and 750 lb. See J.E.T. Rogers, *A history of agriculture and prices in England from 1259–1793* (Clarendon Press), vol. 5 (1887).
32 Rogers, *Agriculture and prices*, vol. 5.
33 Houghton, *Collection*, 5: 110 (7 September 1694).
34 Defoe, *Tour*, pp. 146–7.
35 Haldane, *Drove roads*, p. 60.
36 F.H. Garner, *The cattle of Britain* (Longmans, Green and Co., 1945), p. 71.
37 Markham, 'English housewife', chapter 6; J. Smith, *England's improvement revived, digested into six books* (1670), p. 176; Houghton, *Collection*, 7: 149 (7 June 1695), 7: 150 (14 June 1695) and 7: 154 (12 July 1695).
38 Trow-Smith, *Livestock husbandry*, vol. 1, pp. 237–8. Sir Cyril Wyche's dairy accounts are HFRC Reading, Norf 14/1/1 and are analysed as appendix 1 of Russell, 'Animal breeding', pp. 407–10.
39 Garner, *Cattle of Britain*, pp. 17–18.
40 G.E. Fussell, 'The size of English cattle in the 18th century', *Agricultural History*, 3 (1929), 160–81.
41 Jacob, *Vade mecum*, pp. 13–15.
42 The table of raw data concerned here is given in Russell, 'Animal breeding', p. 215.
43 These figures are in agreement with those derived by Trow-Smith for cattle breeds in the late eighteenth century. See *Livestock husbandry*, vol. 2, pp. 84–120, *passim*.
44 W. Ellis, *The practical farmer or the Hertfordshire husbandman* (London, 1738), part 1, pp. 105–10.
45 An experiment quoted by Young showed that the cream of Longhorn cattle would yield about 1.3 lb of butter per quart of cream while Shorthorns would give only 0.94 lb per quart. See A. Young, *A six months tour through the north of England* (London, 1770), vol. 2, pp. 134–5.
46 Hale, *Compleat body*, pp. 216, 556–7.
47 'Country gentleman', *Complete grazier*, pp. 4–6.
48 Peak lactation occurred in the summer months. In two cases this was in May and June and in one case in September. The period from January to March represented the low point in production when most cows were dried off. See A. Young, 'A tour through Sussex', *Annals of Agriculture*, 22 (1794), 600–2; ESRO, Schiffner archive 3571; and KAO, U471 Waldershare Park MSS A45.
49 W.B. Mercer, 'William Tompson: a record of georgian farming', *Min. Agric. Journal*, 45 (1939), 1125–32.
50 Markham, 'Cheap and good husbandry', p. 88.
51 The full data for this is given as appendix 2 of Russell, 'Animal breeding', pp. 411–15.

52 S. Hartlib, *His legacie or an enlargement of the discourse of husbandry used in Brabant and Flanders* (London, 1651), p. 96.

53 J. Houghton, *A collection of letters* 2: 6 (16 June 1684), pp. 154–61.

54 Lisle quoted a noted Leicestershire dairywoman of the early eighteenth century to the same effect. See E. Lisle, *Observations in husbandry . . .* (London, 1757), pp. 273–4. Elsewhere Houghton again expressed his disbelief that external points could possibly be a good guide to milk potential. See Houghton, *Collection*, 5: 111 (14 September 1694).

55 Lisle's informants suggested that in Hampshire at the end of the seventeenth century they seldom kept their milk cows more than 12 years. Lisle, *Observations*, p. 274. (Modern dairy cows are seldom kept for more than four lactations.)

56 Houghton, *Collection*, 5: 107 (17 August 1694) and 5: 108 (24 August 1694).

57 Houghton, *Collection*, 5: 113 (28 September 1694). In the seventeenth century the average herd size among farmers, yeomen and gentlemen in Lancashire was perhaps 10–12 cows. See O. Ashmore, 'Inventories as a source of local history 2: farmers', *Amateur Historian*, 4 (1959), 186–95.

58 An analysis of the income of Sarah Fell from renting her bulls out to small cattle owners shows that the most active season for bulling was July and August, so that small owners did not manage to bull their cows in May and June, which were in fact months of minimal bull usage. By contrast, the months of February to April represented a small peak of demand. Data derived from N. Penny (ed.), *The household account book of Sarah Fell of Swarthmoor Hall, Lancashire* (Cambridge, 1920).

59 Houghton, *Collection*, 5: 106 (10 August 1694).

60 Lisle also stated that two years old was considered the best age for a breeding bull but noted that hillish breeds required special attention to bring them to sexual maturity at that age. See Lisle, *Observations*, pp. 166–7, 279–80.

61 Rogers, *Agriculture and prices*, vol. 5, p. 332; KAO, U951 Knatchbull MSS A4.

62 Lisle, *Observations*, p. 278.

63 Lisle, *Observations*, p. 279.

64 Lisle, *Observations*, p. 285.

65 Lisle, *Observations*, pp. 274–6.

66 Haldane, *Drove roads*, p. 60.

67 Bailey and Culley agreed that the Chillingham cattle gave excellent beef and that they would probably feed well. See Bailey and Culley, *Agriculture of Northumberland*, p. 17.

68 Thirsk, *Agrarian history*, vol. 4, pp. 83–9. In the forest of Pendle in the seventeenth century the importance of the dairy trade is obvious from inventories. See M. Brigg, 'The forest of Pendle in the seventeenth century', *Trans. Hist. Soc. Lancs. and Cheshire*, 113 (1962), 83–4.

69 J. Harland (ed.), 'The house and farm accounts of the Shuttleworths of Gawthorpe Hall in the county of Lancaster, Sept. 1582–Oct. 1621', *Chetham Society Publications*, 35, 41, 43, 46 (1856–8).

70 The price of all cattle, based on an index value 100 in 1450–99, rose from 391 in 1583 to 487 in 1621, a price rise of 25%. See Thirsk, *Agrarian history*, vol. 4, p. 837.

71 Defoe, *Tour*, pp. 225, 408–9.

72 For instance, George Shakerley of Gwenilt, Wales, had bought a bull calf and accompanying heifer in Lancashire in 1728 for his cousin, Sir William Williams, in Shropshire. He had heard that larger and better cattle were to be

had 'in the further part of Lancashire' and had directions for the acquisition of such animals should Williams desire them. See D. Howell, 'Landlords and estate management in Wales', in J. Thirsk (ed.), *Agrarian history of England and Wales, vol. 5, 1640–1750* (Cambridge University Press, 1985), pp. 252–97.

73 Laurence, *System of agriculture*, pp. 130–1.

74 Mortimer, *Husbandry*, pp. 166–8.

75 Although in the 1680s only the Hertfordshire men had milk-fed calves. See Houghton, *A Collection of letters*, 2 (June 1684), pp. 164–5.

76 Laurence evidently copied Houghton's account of the practice of the Essex veal men and may have learned of the Lancashire nurse cows from the same source.

77 Ellis, *The Modern husbandman*, vol. 3, pp. 93–6, 152–3; *Practical farmer*, part 1, pp. 105–10.

78 For instance, see Anon., 'Housing of great cattle'.

79 Hale, *Compleat body*, pp. 209–16.

80 Any excess valuation of the bull over the cow can usually be explained by supposing the bull to have been fatted. Some accounts show comparable sales of bulls and milk cows in the same year. For instance, KAO, U386 Darrel MSS A2; Berks RO, D/Ewe A2; Berks RO, D/EPb E7; HFRC Reading, Not 4/1/1, Sal 5/1/1, Ken 13/1/1.

81 Two favourable reports concerning this practice appeared in Dossie's *Memoirs*. See R. Dossie, *Memoirs of agriculture and other oeconomical arts* (London, 1782), vol. 3, pp. 299–305.

82 Hale, *Compleat body*, p. 211.

83 Hale, *Compleat body*, pp. 556–7.

84 'Country gentleman', *Complete grazier*, pp. 11–13.

85 Anon. 'Extraordinary size of neat cattle'.

86 HFRC Reading, Sal 5/1/1.

87 Berks RO, D/ED E48C.

88 LAO, Dixon MSS 4/1.

89 Young, *East of England*, vol. 1, pp. 183–4.

90 Berks RO, D/EBu A5/1–2.

91 Young, *East of England*, vol. 1, pp. 110–34.

92 Marshall, *Midland counties*, vol. 1, pp. 295–338.

93 R. Wallace and J.A. Scott-Watson, *Farm livestock of Great Britain* (Edinburgh, 1923), p. 97; Pitt, *Agriculture of Leicestershire*, p. 218.

94 Marshall, *Midland counties*, vol. 1, p. 318.

95 Young, *East of England*, vol. 1, pp. 50–1, 75, 90, 162–3, 170–1; Young, *North of England*, vol. 2, pp. 134–6 and vol. 3, p. 206.

96 Marshall, *Midland counties*, vol. 1, p. 320. *Old Comely* died in 1791 when about 26 years old. Pieces of her carcass were then used as demonstration specimens. Even at that age she had four inches of solid sirloin fat. Clearly Webster's stock were already animals with a prodigious capacity for putting on weight. See H.C. Pawson, *Robert Bakewell: pioneer livestock breeder* (Crosby Lockwood, 1957), pp. 53–5.

97 Marshall, *Midland counties*, vol. 1, pp. 318–21.

98 Marshall, *Midland counties*, vol. 1, pp. 335, 338. John Monk also noticed the youth of Leicester bulls in *Agriculture of Leicestershire*, pp. 27–30. Marshall also noticed the phenomenon for which the very young bull had probably first been used in Lancashire, the bringing of all the herd into milk at the same time. In Warwickshire, where most of the Midland dairies were located, most of the

cows were dried off together, on one day in mid-December. See Marshall, *Midland counties*, vol. 1, pp. 346–9.

99 J.H. Campbell, 'Answers to queries relating to the agriculture of Lancashire', *Annals of Agriculture*, 20 (1793), 121–2. However, Campbell was a firm advocate of Herefords and generally held a poor view of Longhorns, so that his views on this breed cannot be regarded as impartial. See Marshall, *Midland counties*, vol. 1, p. 489; and R.W. Dickson and W. Stevenson, *General view of the agriculture of the county of Lancashire* (London, 1815), pp. 541–3.

100 Penny, *Accounts of Sarah Fell*.

101 KAO, U593 Tylden MSS A5.

102 C.S. Orwin and C.S. Orwin, *The open fields* (3rd edn, Clarendon Press, 1967), pp. 127–8.

103 HFRC Reading, Ber 43/2/2.

104 Berks RO, D/P 93/1/1.

105 Professor Rupert Hall has drawn my attention to the fact that W.E. Tate found bull purchases and sales common in eighteenth-century parish records and that the bulls were generally sold for a little less than their purchase price.

106 Berks RO, D/P 48/12/1.

107 Although Oxfordshire was one of the last counties to be enclosed, its agriculture in the seventeenth and eighteenth centuries was buoyant enough to provide considerable supplies for the London market. Livestock holdings also showed a dramatic increase in the county from the early seventeenth century to the early eighteenth century. All this correlates well with a potentially advanced approach to livestock breeding. See M.A. Havindon, 'Agricultural progress in open-field Oxfordshire', in E.L. Jones (ed.), *Agriculture and economic growth in England, 1650–1815* (Methuen, 1967), pp. 66–7, 74–5.

108 J.H. Fearon, 'Parish accounts for the "town" of Bodicote, Oxfordshire, 1700–1822', *Banbury Hist. Soc.*, 12 (1975).

109 See KAO, U386 Darrel MSS A2; 'Farm accounts of Sir Edward Filmer, Mich. 1738–Mich. 1739', in Melling, *Kentish sources*, vol. 3 (1961), pp. 50–3; KAO, U1127 Smith Masters MSS A9; ESRO, Hooke MSS 16/7 and 16/22; and HFRC Reading, Sal 5/1/1. In the 1760s and 1770s 1*s* and 6*d* remained common bulling fees in Kent but in the 1780s, although some records still show 1*s* as the fee, 2*s* 6*d* was now reasonably common. See KAO, U951 Knatchbull MSS A21; and KAO, U1776 Hussey MSS E2. In Berkshire, however, fees in the 1760s and 1770s were generally 6*d* rising to 1*s* by the 1790s. See Berks RO, D/ESv(M) E8; D/EHr E2; and D/ESv(M) F52, 54, 55, 56, 70, 74. As late as 1806, in Westmorland, the standard fee still seemed to be only 2*s* although this county was one of the source areas for improved Longhorns. The same fee was charged in Derbyshire in 1795 and Kames estimated that 2*s* 6*d* was the standard fee. See HFRC Reading, microfilm P242 and P434; and Home (Lord Kames), *The gentleman farmer . . .* (London, 1776), p. 39.

110 However, by the 1780s the situation had slightly changed as the spirit of livestock improvement spread. At Shardeloes in Buckinghamshire in 1783 and 1785, 6 and 9 guineas respectively were paid for bulls. Cows in this account were generally bought for about £6. See Bucks RO, Drake MSS D/DR/2/114.

111 For instance, *Old Fine Cray* and *Young Fine Cray* in KAO, U771 A18; *Old* and *Young Darking* in LAO, Dixon MSS 4/1; and *Old* and *Young Nancy* in KAO, U593, Tylden MSS A3.

7 Breeding sheep: mutton displaces wool

1 Thirsk, *Agrarian history*, vol. 4, p. 53; Trow-Smith, *Livestock husbandry*, vol. 1, pp. 193–4; Tusser, *Five hundred points of good husbandry*, pp. 84–5; Harrison, 'Description of England', book 3, chapter 3.

2 For details of the medieval fleece, see M.L. Ryder, 'British medieval sheep and their wool types', in D.W. Crossley (ed.), *Medieval industry* (Council for British Archaeology, Report no. 40, 1981), pp. 16–27. For historical records of medieval fleece weights, see Russell, 'Animal breeding', pp. 420–31. Essentially primitive fleece characters persisted well into the seventeenth century. See M.L. Ryder, 'Wool in 1641', *Journal of the Bradford Textile Society for 1969–70* (1971), 36–41.

3 Kerridge, *Agricultural revolution* (George Allen and Unwin, 1967).

4 From 1650 onwards the most dramatic changes in the Midlands were produced by improved livestock grazing on permanent pasture rather than by the novel use of long leys in alternate husbandry. See Broad, 'Alternate husbandry'.

5 Welsh sheep were recorded in Essex inventories in the seventeenth century. See F.W. Steer, *Farm and cottage inventories of mid-Essex, 1635–1749* (Essex Record Office, 1950).

6 Thirsk and Cooper, *Economic documents*, pp. 414–15, 'John Aubrey's list of Wiltshire fairs'. Western sheep were being sold at Leighton Buzzard from estates at Chichely in south-east Northamptonshire in 1665. See Bucks RO, Chester MSS D/C/4/8.

7 G. Markham, 'Cheap and good husbandry', pp. 107–21.

8 R. Plot, *The natural history of Staffordshire* . . . (Oxford, 1686), chapter 7, para. 61; Hervey, *Breviary of Suffolk*, p. 38; R. Carew, *The survey of Cornwall* (1602), book 1 (Adams and Dart reprint, London, 1969), pp. 106–7; R.H. Tawney and E. Power (eds.), *Tudor economic documents* (London, 1924), vol. 1, p. 194.

9 Thirsk and Cooper, *Economic documents*, pp. 191–3; J.H. Munro, 'Wool price schedules and the quality of English wools in the later middle ages *c.* 1270–1499', *Textile History*, 9 (1978), 118–69.

10 The origin of the longer-woolled pasture fleeces of the Midlands and eastern counties has been the subject of historical controversy. Ryder believes, on grounds of biological difference between modern longwools and smaller breeds, that the ancestral longwool was always a different biological variety. By contrast, Bowden thinks, on grounds of a rise in volume of wool traded and a deterioration in its quality in the sixteenth century, that the sheep walks and enclosures of the fifteenth and sixteenth centuries produced better nutrition, which allowed the development of the larger pasture variety from the arable type. This argument is essentially environmental and sees the divergence merely as a consequence of better feeding. See, for instance, P.J. Bowden, *The wool trade in Tudor and Stuart England* (Macmillan, 1962) and 'Wool supply and the woollen industry', *EHR*, 2nd series, 9 (1956–7), 44–58; and M.L. Ryder, 'A history of sheep breeds in Britain', *AHR*, 12 (1964), 65–82.

11 Trow-Smith, *Livestock husbandry*, vol. 2, pp. 36–41, 121–53; G.E. Fussell, 'Animal husbandry in 18th-century England, part 2', *Agricultural History*, 11 (1937), 189–214.

12 M.L. Ryder in various places, but perhaps most cogently in 'Sheep breeds in Britain', pp. 1–12, 65–82; 'The origin and history of British breeds of sheep', *Ark*, 3 (June 1976), 166–70; and 'Medieval sheep and wool types', *AHR*, 32 (1984), 14–28.

13 Kerridge, *Agricultural revolution*, chapter 9; Kerridge, 'Wool growing and wool textiles in medieval and early modern times', in J.G. Jenkins (ed.), *The wool textile industry in Great Britain* (Routledge, 1972), pp. 19–33; R.M. Hartwell, 'A revolution in the character and destiny of British wools', in N.B. Harte and K.G. Ponting (eds.), *Textile history and economic history* (Manchester University Press, 1973), pp. 320–38.

14 M.L. Ryder, 'A survey of European primitive breeds of sheep', *Ann. Génét. Sél. Anim.*, 13 (1981), 381–418.

15 T.A. Knight, 'An account of Herefordshire breeds of sheep, cattle, horses and hogs', in *Communications to the Board of Agriculture*, vol. 2 (1800), pp. 174–5; W. Marshall, *The rural economy of Gloucestershire* (Gloucester, 1789), vol. 2, pp. 233–8; W. Redhead, R. Laing and W. Marshall, *Observations respecting the different breeds of sheep in some of the principal counties in England* (Edinburgh, 1792), pp. 23–4.

16 J.A. Perkins, *Sheep farming in 18th- and 19th-century Lincolnshire* (The Society for Lincolnshire History and Archaeology, 1977), p. 9; D. Defoe, *A plan of the English commerce* (London, 1728), p. 160; G. Turner, *General view of the agriculture of the county of Gloucestershire* (London, 1794), p. 9.

17 Markham, 'Cheap and good husbandry', pp. 107–8; Defoe, *Tour*, pp. 408–9, 413; Defoe, *English commerce*, pp. 157–8, 160–1; J. Munn, *Observations on British wool . . .* (London, 1738), p. 314; Ellis, *Practical farmer*, pp. 116–17; W. Ellis, *A compleat system of experienced improvements . . .* (London, 1749), book 1, chapter 4; Kerridge, 'Wool growing', p. 26.

18 R. Proctor-Aldernon, 'Devonshire sheep', *Annals of Agriculture*, 17 (1792), 299–300; J. Luccock, *An essay on wool . . .* (London, 1809), pp. 187, 286–7; Ryder, 'Sheep breeds in Britain', p. 80; W. Marshall, *The rural economy of the southern counties* (London, 1798), vol. 1, pp. 378–9.

19 Ryder, 'Sheep breeds in Britain', p. 65; Bailey and Culley, *Agriculture of Northumberland*, pp. 19–21; Trow-Smith, *Livestock husbandry*, vol. 2, pp. 272–3; Pawson, *Robert Bakewell*, pp. 66–7; Luccock, *Essay on wool*, p. 331.

20 R. Fraser, *General view of the agriculture of the county of Devon* (London, 1794), pp. 34, 36–7; Claridge, *Agriculture of Dorset*, pp. 8–11; J. Billingsley, *General view of the agriculture of the county of Somerset* (London, 1794), pp. 82–3, and 106; W. Marshall, *The rural economy of the West of England* (London, 1796), vol. 1, pp. 260–3.

21 T. Davis, *General view of the agriculture of the county of Wiltshire* (London, 1794), pp. 20–2; A. Driver and W. Driver, *General view of the agriculture of the county of Hampshire* (London, 1794), p. 23; Rev. Mr Warner, *General view of the agriculture of the Isle of Wight* (London, 1794), pp. 53–4.

22 Lisle, *Observations*, p. 424; A. Young, 'Miscellanies in Hampshire', *Annals of Agriculture*, 23 (1795), 167; Trow-Smith, *Livestock husbandry*, vol. 2, p. 129; Ellis, *Compleat system*, book 1, chapter 6.

23 C. Vancouver, *General view of the agriculture of the county of Cambridge* (London, 1794), pp. 205–6; Luccock, *Essay on wool*, pp. 238–42, 253–4; Rev. D. Howlett, 'Answers to queries', *Annals of Agriculture*, 15 (1791), 263; Defoe, *English commerce*, p. 160; N. Kent, *General view of the agriculture of the county of Norfolk* (London, 1794), pp. 32–4; Young, *Agriculture of Suffolk*, pp. 33–6.

24 Marshall, *Midland counties*, vol. 2, minute 60.

25 J. Wedge, *General view of the agriculture of the county of Warwickshire* (London, 1794), p. 31.

26 Defoe, *English commerce*, p. 160.

27 Munn, *British wool*, p. 2.
28 Munn, *British wool*, p. 4; J. Smith, *A review of the manufacturer's complaints against the wool growers* . . . (London, 1753), pp. 17–18.
29 Knight, 'Herefordshire breeds', p. 174.
30 E.L. Jones, 'Hereford cattle and Ryeland sheep, economic aspects of breed changes, 1780–1870', *Trans. Woolhope Naturalists Field Club*, 38 (1965), 36–48.
31 Turner, *Agriculture of Gloucestershire*, pp. 9–11; Marshall, *Gloucestershire*, vol. 1, pp. 233–7; J. De Lacey Mann, *Cloth industry in the West of England, 1640–1880* (Oxford University Press, 1971), pp. 257–8.
32 Perkins, *Sheep farming in Lincolnshire*, pp. 33–5; Young, *North of England*, vol. 2, p. 17; Marshall, *Yorkshire*, vol. 2, pp. 260–1; W. Strickland, 'Reply to queries', *Annals of Agriculture*, 15 (1791), 313–14; I. Leatham, *General view of the agriculture of the East Riding of Yorkshire* (London, 1794), pp. 51–2; Luccock, *Essay on wool*, p. 325.
33 See the first editions of the *Board of Agriculture County Reports* for counties with a significant amount of lowland grazing and pasture.
34 Brown, *Agriculture of Derbyshire*, pp. 25–6; Luccock, *Essay on wool*, pp. 311–12, 314–15, 298–300; R. Lowe, *General view of the agriculture of the county of Nottingham* (London, 1794), pp. 17–100 *passim*; W. Pitt, 'Reply to queries', *Annals of Agriculture*, 15 (1791), 319; W. Pitt, *General view of the agriculture of the county of Staffordshire* (London, 1794), pp. 52–61; Young, *North of England*, vol. 4, p. 339; Thomas, *Agriculture of Worcestershire*, p. 11.
35 Young, *East of England*, vol. 1, pp. 459–60, 467–8.
36 Bailey and Culley, *Agriculture of Northumberland*, pp. 19–21.
37 Redhead *et al.*, *Breeds of sheep*, p. 27; Sir J. Sinclair, 'Description of the Cheviot and analysis of a Cheviot sheep farm', appendix 2 to Redhead *et al.*, pp. 65–8.
38 Carlyle, 'Breeds of sheep in Scotland'.
39 Imported Wiltshire and Dorset sheep were commonly recorded in mid-eighteenth-century Kent farm accounts. For instance, KAO, U46/47 Streatfield MSS A1 (1765), U951 Knatchbull MSS F 18/1 and 18/2 (1731–44), U269 Sackville MSS A302 (1730), U1127 Smith Masters MSS A9 (1740s). Also in West Sussex and Buckinghamshire. See E. Turner (ed.), 'The farming diary of Richard Stapley of Twineham, Sussex, 1682–1724', *Sussex Arch. Collections*, 2 (1849), 102–28; and HFRC Reading, microfilm P321, Buc 5/2.
40 C. Vancouver, *General view of the agriculture of the county of Essex* (London, 1795), pp. 128–9; Vancouver, *Agriculture of Cambridge*, pp. 205–6; Howlett, 'Answers to queries', pp. 263–4.
41 De Lacey Mann, *Cloth industry*, p. 258.
42 Lisle, *Observations*, p. 307.
43 Davis, *Agriculture of Wiltshire*, pp. 22–3.
44 Marshall, *Norfolk*, vol. 1, pp. 363–4; Marshall, *Midland counties*, vol. 1, pp. 412–13.
45 Bishton, *Agriculture of Shropshire* (London, 1794), pp. 10–11; Redhead *et al.*, *Breeds of sheep*, pp. 16–18; Pitt, *Agriculture of Staffordshire*, p. 56; Knight, 'Herefordshire breeds', p. 172.
46 T. Wedge, *General view of the agriculture of the county of Cheshire* (London, 1794), p. 28; Luccock, *Essay on wool*, p. 315; Lowe, *Agriculture of Nottingham*, pp. 17–160 *passim*; Pitt, *Agriculture of Staffordshire*, pp. 52–5.
47 Munn, *British wool*, preface; G. White, *The natural history of Selborne* (1789)

(Oxford, 1974), pp. 167–8; A. Young, *General view of the agriculture of the county of Sussex* (London, 1794), p. 57.

48 E.L. Jones, 'The entry of the Southdown sheep into the Wessex chalklands', *J. Agric. Club of the University of Reading* (1960), 38–40; A. Young, 'Experiments on the introduction of the Southdown sheep into Suffolk', *Annals of Agriculture*, 15 (1791), 286–310.

49 Brown, *Agriculture of Derbyshire*, pp. 24–5; Redhead *et al.*, *Breeds of sheep*, p. 9; Rennie, Brown and Shirreff, *General view of the agriculture of the West Riding of Yorkshire* (London, 1794), p. 32.

50 J. Tuke, *Agriculture of the North Riding* (London, 1794), pp. 65–6; Marshall, *Yorkshire*, vol. 2, pp. 221–2; Redhead *et al.*, *Breeds of sheep*, pp. 5–6; A. Pringle, *General view of the agriculture of the county of Westmorland* (London, 1794), pp. 24, 27; Bailey and Culley, *Agriculture of Northumberland*, p. 22; *General view of the agriculture of the county of Cumberland* (London, 1794), p. 16. Grainger suggested that in Weardale in Durham there was still an ancient breed of partly horned fine-wools with mottled faces, which were probably a remnant of an old fallow, field or common sheep, similar to those still found at that date reasonably extensively further south. See J. Grainger, *General view of the agriculture of the county of Durham* (London, 1794), pp. 36–7.

51 Bailey and Culley, *Agriculture of Cumberland*, pp. 14–16; Redhead *et al.*, *Breeds of sheep*, pp. 6–7; C.M.L. Bouch, G.P. Jones and R.W. Brunskill, *Short history of the Lake counties, 1500–1830* (Manchester University Press, 1961), pp. 347–8.

52 Rogers, *Agriculture and prices*, vol. 5 (1887).

53 National Sheep Breeders Association, *British sheep* (London, 1968).

54 ESRO, Dunn MSS 45/2/89.

55 K.J. Allison, 'Flock management in the sixteenth and seventeenth centuries', *EHR*, 2nd series, 11 (1958–9), 105.

56 HFRC Reading, Ken 19/1/1.

57 Fussell initiated a survey and abstract of the important literature, giving carcass and fleece weights of eighteenth-century sheep half a century ago. See G.E. Fussell and C. Goodman, '18th-century estimates of British sheep and wool production', *Agricultural History*, 4 (1930), 131–51.

58 Young believed that Southdown wethers generally averaged 16–20 lb per quarter (64–80 lb carcass weight), while the data here relate to the leading Southdown flock and its derivatives. See Young, *Agriculture of Sussex*, p. 47.

59 Davis, *Agriculture of Wiltshire*, pp. 22–3.

60 Young, *East of England*, vol. 1, pp. 183–4.

61 Young reported outstanding Lincoln longwools in the early nineteenth century as of the same order of size. See A. Young, *General view of agriculture of the county of Lincolnshire* (London, 1813), pp. 344–9.

62 House of Commons parliamentary report, *Report from the committee upon the petitions relating to the false winding of wool, and the marking of sheep with pitch and tar* (London, 1752), p. 7.

63 Perkins, *Sheep farming in Lincolnshire*, pp. 19–40.

64 Stone, *Lincoln*, p. 61; Young, *Agriculture of Lincolnshire*, pp. 411–12.

65 Perkins, *Sheep farming in Lincolnshire*, quoting *Lincoln, Rutland and Stamford Mercury*, 12 June 1846.

66 Fitzherbert, *Boke of husbandry* and Tusser, *Five hundred points of good husbandry*.

67 Tusser, *Five hundred points of good husbandry*, p. 102.

68 T. Mouffet, *Health's improvement . . . or rules . . . for preparing all sorts of food*

(London, 1655), pp. 63–4. From the mid-sixteenth century meat production became much more important for farmers in Norfolk, who supplied both the Norwich and London markets. The market was not only for cull crones and pucks but also for fat wethers. See Allison, 'Flock management', pp. 108–9.

69 Markham, 'Cheap and good husbandry', pp. 107–21; Trow-Smith, *Livestock husbandry*, vol. 1, pp. 241–6.

70 C.B. Robinson, 'Rural economy in Yorkshire in 1641', *Surtees Society Publications*, 33 (1857).

71 Trow-Smith, *Livestock husbandry*, vol. 1, pp. 245–6; Bowden, *Wool trade*, pp. 22–4; Ryder, 'Wool in 1641'.

72 Ryder, 'Wool in 1641', p. 36.

73 Robinson, 'John Best's inventory 1668' in 'Rural economy in Yorkshire'.

74 W.H. Long, 'Regional farming in 17th-century Yorkshire', *AHR*, 8 (1960), 103–24.

75 In Norfolk a market for fat lamb had developed in the mid-seventeenth century and was well developed by 1700. See Allison, 'Flock management', p. 109.

76 Robinson, 'Rural economy in Yorkshire', pp. 1–13, *passim*; Trow-Smith, *Livestock husbandry*, vol. 1, pp. 241–6.

77 Hartlib, *His legacie or an enlargement of the discourse of husbandry used in Brabant and Flanders*, pp. 97–8.

78 J. Beale, 'Some inquiries and suggestions concerning salt for domestic uses, and concerning sheep, to preserve and improve the race', *Phil. Trans. Royal Society*, 9 (1674), 48–53.

79 Bowden, *Wool trade*, p. 20.

80 M.E. Finch, 'The wealth of five Northants families, 1540–1640', *Northants Record Society*, 19 (1956), 40.

81 Finch, 'Five Northants families', quoting BM Add. MSS 25079, fol. 55.

82 *Pickering's statutes*, vol. 3, pp. 95–6.

83 *Pickering's statutes*, vol. 6, pp. 234–5.

84 *APC*, vol. 33 (1613–26), 27 March 1614.

85 Lodge, *Account book of a Kentish estate, passim*.

86 HFRC Reading, Ken 19/1/1.

87 HMC, 'Manuscripts of T.B. Clarke Thornhill Esq. of Rushton Hall, Northants' (London, 1904).

88 A point well understood by some of Lisle's colleagues. See Lisle, *Observations*, p. 306.

89 Nichols, *Livestock improvement*, chapter 15; Carlyle, 'Breeds of sheep in Scotland'; Trow-Smith, *Livestock husbandry*, vol. 2, *passim*; Hammond, *Farm animals*, pp. 241–2.

90 Maxwell found that in England, as in Scotland, gentlemen generally knew nothing of sheep husbandry, leaving such matters entirely to their shepherds. See G.C. Maxwell, *Observations on . . . growing wool in Scotland* (Edinburgh, 1756), p. 18.

91 Such views were implicit in the discussion of sheep found in all the following: Mortimer, *Husbandry*; Jacob, *Vade mecum*; and Bradley, *Gentleman and farmer's guide*.

92 J. Anderson, *Essays relating to agriculture and rural affairs . . .* (Edinburgh, 1775), pp. 305–33. In the early nineteenth century Luccock believed that wool quality was a consequence of breed, but the geologist, Robert Bakewell, still thought the fleece form was completely under environmental control. See

Luccock, *Essay on wool*, section 1, *passim*; and R. Bakewell, *Observations on the influence of soil and climate upon wool* . . . (London, 1808).

93 Biographical and analytical details about Bakewell appear in many places but the most comprehensive accounts of the man and his work are to be found in the following: W. Houseman, 'Robert Bakewell', *J. Royal Agric. Soc. of England*, new series, 5 (1894), 1–31; Pawson, *Robert Bakewell*; Trow-Smith, *Livestock husbandry*, vol. 2, chapter 2; J.A.S. Watson, 'Bakewell's legacy', *J. Royal Agric. Soc. of England*, 89 (1928), 22–32. Some of the major activities of his life-long disciple, George Culley, have been outlined in D.J. Rowe, 'The Culleys, Northumberland farmers 1767–1813', *AHR*, 19 (1971), 156–74; S. Macdonald, 'The role of George Culley of Fenton in the development of Northumberland agriculture', *Arch. Aeliana*, 5th series, 3 (1974), 131–41. The most reliable contemporary sources about him are the following: Young, *East of England*, vol. 1, pp. 110–34; Young, 'Ten days tour to Mr Bakewell's', pp. 452–502; A. Young, *On the husbandry of three celebrated farmers* . . . (London, 1811); J. Boys and J. Ellman, 'Agricultural minutes taken . . . Rutland, Leicester', *Annals of Agriculture*, 19 (1793), 72–146; A. Young, 'A month's tour to Northamptonshire, Leicestershire . . .', *Annals of Agriculture*, 16 (1791), 480–607; J.H., 'A report on a visit to Mr Bakewell's farm', *Gentleman's Magazine*, 63 (1793), 792–5; Monk, *Agriculture of Leicestershire*, pp. 27–30; R. Parkinson, *The experienced farmer* . . . (London, 1798), vol. 1, pp. 7–13; Redhead *et al.*, *Breeds of sheep*, pp. 33–9; Anon., 'Obituary of Robert Bakewell', *Gentleman's Magazine*, 65 (1795), 969–71; Marshall, *Midland counties*, vol. 1, pp. 295–338, *passim*; G. Culley, 'An account of the pedigree of Mr Culley's tups', *Annals of Agriculture*, 28 (1797).

94 These are drawn from the primary and secondary sources listed *in extenso* in note 93.

95 Rowe, 'The Culleys', pp. 160–1.

96 Marshall, *Midland counties*, vol. 1, pp. 434–5; A. Young, 'A tour in Sussex', *Annals of Agriculture*, 11 (1789), 213.

97 Marshall, *Midland counties*, vol. 1, p. 301; Young, 'Tour in Sussex', pp. 215–18; Young, 'Tour through Sussex', pp. 228–9, 241–2.

98 Redhead *et al.*, *Breeds of sheep*, pp. 34–6; Young, *East of England*, vol. 1, p. 119; Young, 'Ten days tour to Mr Bakewell's', p. 483; Boys and Ellman, 'Agricultural minutes', pp. 128–30; Monk, *Agriculture of Leicestershire*, pp. 27–8; A. Young, 'Some farming notes in Essex, Kent and Sussex', *Annals of Agriculture*, 20 (1793), 261–2.

99 A. Young, 'Experiments in weighing fatting cattle alive', *Annals of Agriculture*, 14 (1790), 140–63.

100 Trow-Smith, *Livestock husbandry*, vol. 2, p. 54.

101 Culley, article in *Annals of Agriculture* (1797), reprinted in Pawson, *Robert Bakewell*, p. 194–5.

102 The small Leicestershire farmer, John Clayton, paid 2 guineas for the hire of a ram in the late 1760s, while a grazier at Beaumont Leys, in Leicestershire, paid 7 and 11 guineas for rams in 1808. See E. Smith, 'A 200-year-old Leicestershire farm account book', *Leicestershire Historian*, 1 (1968), 76–81; HFRC Reading, Lei 4/1/1.

103 Young, *East of England*, vol. 1, pp. 116–17; Young, *Celebrated farmers*, p. 9.

104 Marshall, *Midland counties*, vol. 1, p. 417; Young, *Celebrated farmers*, p. 5; Perkins, *Sheep farming in Lincolnshire*, p. 37. He also points out that several

features of the practice of the Dishley Society also seem to have started in Lincolnshire.

105 Young, 'Tour in Sussex'; Young, 'Tour through Sussex'. Sums of 7s and 15s were paid for the hire of rams by one grazier, who also hired out rams to others. Pawson stated that he had evidence (which he did not specify) that rams were hired on Romney Marsh as early as 1706. See KAO U301 Miscellaneous deeds E6; and Pawson, *Robert Bakewell*, p. 70.

106 In the 1790s Codd of Glentworth Heath was still selecting rams and ewes carefully according to fleece quality. From the 1750s to the 1790s the Dixon family selected their ewes and rams for 'skin' or fleece quality, although their detailed records show an attempt to 'nick' the good and poorer-fleeced rams and ewes so that they compensated each other, rather than selecting strongly in one direction for an improved fleece. Nicking was also common practice among the Midland ram breeders and graziers in the 1780s. See Redhead *et al.*, *Breeds of sheep*, p. 43; LAO Dixon 4/1; and Marshall, *Midland counties*, vol. 1, p. 422.

107 House of Commons parliamentary report, *Report of the committee to whom the petition of several gentlemen . . . breeders and feeders of sheep in the county of Lincoln . . .* (London, 1755), pp. 5–9. Prices for rams generally in the eighteenth century were less than £2. For instance, see Bucks RO, D/DR/2/112, D/MH 36/ 2; KAO, U386 A2, U301 E6, U908 E6, U1127 A9; Berks RO, D/EPb E14, D/Ewe A2; and HFRC Reading, Dev 3/2/12, Sal 5/1/1. But more was paid on occasion: for instance, £2 2s in 1784 in Kent (KAO, U769 E1); up to £6 6s on the Berkshire Downs late in the eighteenth century (Berks RO D/EX 62/2); and £2 13s in Yorkshire in 1754. See Ralph Ward's journal in C.E. Whiting (ed.), 'Two Yorkshire diaries . . .', *Yorks. Arch. Soc. Record Series*, 117 (1952).

108 Pennant, *British zoology*, vol. 1, p. 25.

109 Anon. 'Extraordinary size of neat cattle'.

110 LAO, Dixon MSS 4/1.

111 LAO, Dixon MSS 4/1. However, in Derbyshire in the 1790s some graziers were putting very few ewes with each ram. The ewes were individually named. See HRFC Reading, microfilm P434.

112 Marshall, *Midland counties*, vol. 1, pp. 381–2.

113 Marshall, *Midland counties*, vol. 1, p. 417.

114 Young, *Celebrated farmers*, p. 4; Marshall, *Midland counties*, vol. 1, pp. 417, 425.

115 The *Gentleman's Magazine* recorded a Robert Bakewell of Dishley as bankrupt in 1776. Culley certainly assisted in setting him up again and was given free use of a lead tup in about 1780 as a consequence. See Pawson, *Robert Bakewell*, pp. 37–9, 194–5. Young and others believed, however, that the low prices which Bakewell stock obtained at that time were a result of the depression caused by the American war. See Young, *Agriculture of Lincolnshire*, p. 407; Culley, 'Account of Mr Culley's tups'; and Pawson, *Robert Bakewell*, p. 194.

116 Young, *Agriculture of Lincolnshire*, p. 407. Young's informant even claimed that a sum of 100 guineas had been obtained for a ram in 1770. This was, however, hearsay. Young himself was told by Bakewell that 5 to 30 guineas per season was charged. See Young, *East of England*, vol. 1, p. 117.

117 Perkins, *Sheep farming in Lincolnshire*, p. 42; Parkinson, *The experienced farmer*, p. 2; Young, *Celebrated farmers*, p. 9; Young, *Agriculture of Lincolnshire*, pp. 352, 403.

118 Macdonald, 'George Culley', pp. 131–2.

119 Culley, article in *Annals of Agriculture*, quoted in Pawson, *Robert Bakewell*, pp. 194–5.

120 A less exalted pure-bred Dishley flock was still being bred in Lincolnshire by a Mr Walker in 1812 at Woolsthorpe. He seems to have derived his breed from Dishley *Old G*, another sire of the 1770s. See Young, *Agriculture of Lincolnshire*, pp. 406–7.

121 He stated to Young that he allowed parent–sibling matings as a matter of course by that date. See Young, 'Ten days tour to Mr Bakewell's', p. 488.

122 The Shorthorn improvers of the late eighteenth and early nineteenth centuries were said to have derived their methods from Bakewell. Charles Colling's method was to select by purchase a group of outstanding foundation animals and inbreed intensely to them in order to perpetuate their properties in their offspring, the classic eighteenth-century reason for inbreeding. The inbreeding coefficient of Colling's stock rose from zero to 40% and Bates then stabilised the coefficient for the breed at about this value. There is no evidence that Bakewell did the same thing with the New Leicester. In the only other closed breed which has been analysed to the same depth, the nineteenth-century Clydesdale, the inbreeding coefficient in the formative 25 years of the breed remained effectively zero and over the 40 years from 1880 to 1920 it only rose to between 5% and 6%, so that inbreeding to fix the type of the breed or individual did not occur. The breed in 1925 was nonetheless uniform in general standard. See S. Wright, 'Mendelian analysis of the pure breeds of livestock 2', *Journal of Heredity*, 14 (1923), 405–22; A. Calder, 'The role of inbreeding in the development of the Clydesdale breed of horses', *Proc. Royal Soc. of Edinburgh*, 47 (1927), 118–40; and Trow-Smith, *Livestock husbandry*, vol. 2, pp. 233–41.

123 Duke of Bedford, 'Four breeds of sheep'.

124 However, it must be stressed that if the New Leicester was significantly faster maturing, this may have been an unfair comparison because they may have spent the last few months at full size, so that there was no growth conversion then. Not surprisingly, what few hints Bakewell gave away about his trials demonstrated the perfection of the Leicester compared with others. For reasons rehearsed by Young, no reliance should be placed upon these data. See Redhead *et al.*, *Breeds of sheep*; Boys and Ellman, 'Agricultural minutes'; and Monk, *Agriculture of Leicestershire*.

125 *Annals of Agriculture*, 23 (1795), 456–7, 26 (1796), 112–37; 25 (1796), 58–62; 24 (1795), 578; 20 (1793), 220–87.

8 Summary and conclusion

1 P.L. Armitage, 'A preliminary description of British cattle from the late 12th to the early 16th century', *Ark*, 7 (1980), 405–13.

2 Marshall, *Midland counties*, vol. 1, p. 300.

Bibliography

Classical sources

Aristotle. *De generatione animalium*, ed. A.L. Peck, Loeb Classical Library, 1943
  Historia animalium, ed. W.W. D'Arcy Thompson, Oxford University Press, 1910
  Historia animalium (books I–III), ed. A.L. Peck, Loeb Classical Library, 1965
Columella. *De re rustica*, ed. E.S. Forster and E.H. Heffner, Loeb Classical Library, 1954
Lloyd, G.E.R. (ed.). *Hippocratic writings*, Penguin, 1978
Plato. *The republic*, ed. D.P. Lee, Penguin, 1955
Pliny. *Natural history*, ed. H. Rackham, Loeb Classical Library, 1961
Varro, M.T. *Res rusticae*, ed. W.D. Hooper and H.B. Ash, Loeb Classical Library, 1960
Xenophon. *Scripta minora*, ed. E.C. Marchant and G.W. Bowersock, Loeb Classical Library, 1968

Manuscript sources

Berkshire Record Office
D/EBu A5/1–2. *Farm accounts of Sir Robert Burdett of Foremark, Derbyshire, 1764–1797, 2 vols.*
D/EBu A10. *Stable account book of Sir Robert Burdett, 1753–1762*
D/ED E48C. *Farming accounts of Ombersley, Worcestershire, 1756–1769*
D/EHr E2. *Farm accounts of Henry Hunter of Swallowfield, Berkshire, 1769–1777*
D/EPb E7 and E14. *Account books for the Coleshill estate, 1726 and 1730–1738*
D/ESv(M) E8. *Account book of T. Stevens in account with Bradfield parsonage, Berkshire, Jan. 1777–Dec. 1779*
D/ESv(M) F52, 54, 55, 56, 70, 74. *Farm accounts of the Stevens family, Bradfield, mixed with personal accounts and diary, 1779–1782, 1791 and 1795*
D/EWe A2. *Throckmorton farm accounts, Berkshire, 1733–1739*
D/EX 62/2. *Farm diary of Thomas Johnson of Hampstead Norris, Berkshire, 1764–1794*
D/P 48/12/1. *Overseer's book for Drayton, Berkshire, 1739–1794*
D/P 93/1/1. *Parish book for Purley, Berkshire*

Buckinghamshire Record Office
Chester MSS D/C/4/8. *Account book with rentals, wages and livestock bought and sold, 1652–1697. Chichely, Bucks.*
  D/C/4/30. *17th-century farm account book*

Drake MSS D/D/A 170. *Farm stock book for Ombersley, Worcestershire, 1761–1801*
   D/DR/2/61. *John Bigg's account of horse purchases in Staffordshire, 1763*
   D/DR/2/112. *Account between William Drake and John Bigg, 1768*
   D/DR/2/114. *Account between William West and William Drake, 1778–1788*
   D/DR/2/118. *Record of sheep killed at Shardeloes*
Earl of Bucks MSS D/MH 36/2. *Farm daybook for 1788 near Wendover, Bucks.*

East Sussex Record Office
Dunn MSS 38/1. *Almanack with farming accounts for an estate at Bexhill, 1678*
   38/7. *Farming almanack of John Roberts of Ticehurst, 1787*
   45/2/89. *Sheep weight account of Walter Roberts 1687*
Glynde Place archives 2937.*Rentals and disbursements of Richard Trevor, Bishop of
   Durham on the Glynde Place estates. Mid-18th century*
Hooke MSS 16/1–16/29. *Estate, farming and personal diary of William Poole, 1747*
Schniffner archive 3570. *Farm accounts for Coombe in Hamsey, 1768–1772*
   3571. *Farm accounts for Coombe in Hamsey, 1796–1836*

Historic Farm Records Collection at the University of Reading
Ber 43/2/2. *Account book of the Rev. John Aldworth, rector of East Lockinge, Berkshire,
   1716–1729*
Buc 11/1/5. *'Home bargain' book of Richard Greville, unknown Buckinghamshire
   location, 1741–1743*
Dev 3/2/12. *Account book of the Wells family of East Allington, Devon, 1762–1782*
Han 10/1/1. *Receipt book of Robert Bristow of Micheldever, Hants., March to May 1726*
Hert 1/1/1. *Farming account of Richard Boor, steward to James, Earl of Salisbury, 1760–
   1761*
Ken 13/1/1. *Farm account book for Tatlingbury farm, Tudeley, Kent, 1744–1758*
Ken 19/1/1. *Farm account book kept by Henry Deedes of Ruffins Hill farm, Burmarsh,
   Kent, 1696–1720*
Lei 4/1/1. *Farm account book of Joshua Ellis senior, Beaumont Leys, Leicestershire,
   1804–1813*
Lei 4/1/2. *Farm account book of Joshua Ellis junior, 1813–1830*
Norf 14/1/1. *Loose accounts of the Manor farm, Hockwold, Norfolk, 1685–1698*
Norf 17/1/1. *Loose accounts, invoices and receipts of Sir Cyril Wyche of Hockwold,
   Norfolk, 1681–1723*
Not 4/1/1. *Farm account book of Mr Baxendon, Collingham, Newark, Notts., 1748–
   1762*
Sal 5/1/1. *Farm account book of Coton Hall farm, Bridgenorth, Shropshire, 1744–1769*
Microfilm P242. *Farm account book from Kirkby Thorpe, Westmorland, 1806–1860*
   P262. *Accounts of William Tompson of Forge farm, Abbot's Bromley, 1750–1790s*
   P321. (Buc 5/2/2 and 5/2/3). *Farm account book of Jonathon Clarke of Quainton,
   Bucks., 1739–1748 and 1777–1784*
   P434. *Farm account book from a location in Derbyshire, 1710–1790*

Kent Archive Office
U23 Culpepper MSS E7. *Inventory of Lady Culpepper's goods at Leeds Castle, Kent,
   November 1710*
U46/47 Streatfield MSS A1. *Estate account book for Kent, 1765–1776*
U145 Faunce MSS A4/1. *Estate account of Mrs Catharine Swift, 1739*

U269 Sackville MSS A45/1 and 2. *Rough farming notes for Knole, early eighteenth century*
A302. *Bills and receipts for the sale and purchase of sheep at Knole and Stoneland, Sussex, 1718–1752*
A413. *Cattle accounts at Milcote, Gloucestershire, 1625–1627*
E226/10 (16). *Agreement for the sale of Gloucestershire wool, 2nd November 1639*
U301 Misc. deeds E6. *Farm accounts of the Pattenson family in the Weald of Kent, 1766–1783*
U386 Darrel MSS A2. *Farm rental and account book for Coleshill, Kent, 1726–1729*
U471 Waldershare Park MSS A14. *Cash book kept by Edward George of receipts and disbursements for the Kentish estates of the Earl of Guildford, 1767–1791*
A29. *Housebook for Waldershare Park, 1753–1765*
A45. *Dairymaid's accounts, 1798–1800*
A60. *Weekly housekeeping account at Waldershare, 1748–1754*
U593 Tylden MSS A3. *Farm account book of Richard Tylden, 1723–1761*
A5. *Farm accounts of the Osborne family, 1683–1692*
U769 Snell MSS E1. *Account book of sheep, dairy and horsework in the Kentish Weald, 1783–1810*
U771 Osborne MSS A18. *Calf and pigeon book, 1706–1707*
U908 Streathfield accounts E6. *Accounts for Delaware farm, 1739–1745*
U951 Knatchbull MSS A4. *Detailed estate accounts at Mersham-le-Hatch, 1670–1684*
A21. *Edward Knatchbull's estate accounts, 1763–1775*
F18/1–2. *Hatch journals of Sir William Knatchbull, 1731–1742 and 1742–1748*
U1127 Smith Masters MSS A7 and 9. *Farm account books for the Camer estate, Meopham, 1715–1743 and 1740–1755*
U1776 Hussey MSS E2. *Farm account book, Kentish Weald, 1778–1786*

Lincoln Archive Office
Dixon MSS 4/1–2. *Farm account books for North Lincolnshire, 1755–1798*
5/2/2. *Stock inventory book for Thornton and Nettleton, 1805–1812*
Massingberd-Mundy MSS MM5/2. *Housekeeping account of Burrell Massingberd, 1710–1718*
MM6/5, nos. 2, 5, 8, 17, 18, 19, 27, 28. *Memoranda relating to sheep and wool sales, 1655–1685*
Miscellaneous deposit 150/1. *Account book of John Hutchinson of Gedney, 1776–1790s*

West Sussex Record Office
Add. MSS 9432. *Bury farm accounts, Hants–Sussex border, 1755–1759*
Miscellaneous papers 1441. *Correspondence between Charles, 6th Duke of Somerset and his steward, 1701–1704. Transcribed by W.E. Thurgood as 'May it please your grace'*
1478. *Pocket book of John Woods of Chilgrove, West Dean*

Manuscript sources, edited, transcribed or collected
Anon. 'The will of Nicholas Blundell 1736', *Trans. Hist. Soc. of Lancs. and Cheshire*, new series, 30 (1915), 261–4
Bagley, J.J. and Tyrer, F. (eds.). 'Blundell's diary, the great diurnall of Nicholas

Blundell of Crosby, 1702–1728', *Record Society of Lancashire and Cheshire*, 110, 112, 114 (1968, 1970, 1972)
Britten, J. (ed.). *John Aubrey's natural history of Wiltshire*, Wiltshire Topographical Society, (London, 1847)
Cust, Lady E. *Records of the Cust family, series 2: the Brownlows of Belton, 1550–1799*, London, 1909
Derham, W. (ed.). *Philosophical letters between the late learned Mr Ray and several of his ingenious correspondents, natives and foreigners. To which are added those of Francis Willoughby esq.*, London, 1718
Dick, O.L. (ed.). *Aubrey's brief lives*, London, 1972
D'Osma, P. 'Explanation and account of my sojourn at the stud at Malmesbury and then at Tutbury' (1576), in C.M. Prior, *The royal studs . . .*, London, 1935, pp. 11–38
Fearon, J.H. 'Parish accounts for the "town" of Bodicote, Oxfordshire, 1700–1822', *Banbury Historical Society Publications*, 12 (1975)
Fussell, G.E. 'Robert Loder's farm accounts, 1610–1620', *Camden Society Publications*, 3rd series, 53 (1936)
Harland, J. (ed.). 'The house and farm accounts of the Shuttleworths of Gawthorpe Hall in the county of Lancaster, Sept. 1582–Oct. 1621', *Publications of the Chetham Society*, 35, 41, 43, 46 (1856–8)
Hervey, Lord F. (ed.). *The breviary of Suffolk by Robert Reyce of 1618*, London, 1902
Hervey, S.H.A. (ed.). *The diary of John Hervey, 1st Earl of Bristol, with extracts from his book of expenses, 1688–1742*, Wells, 1894
HMC *11th Report appendix part IV*, 'Manuscripts of the Marquess Townsend', London, 1887
    12th Report appendix part IV, 'Manuscripts of his grace the Duke of Rutland, preserved at Belvoir Castle', 4 vols., London, 1888–1905
    13th Report appendix, 'Manuscripts of the Duke of Portland at Welbeck', 10 vols., London, 1891–
    14th Report appendix part VII, 'Manuscripts of the city of Lincoln', London, 1895
    Report on MSS in various collections, vol. 3, 'Manuscripts of T.B. Clarke Thornhill Esq. of Rushton Hall, Northants', London, 1904
    Report on the manuscripts of the Earl of Ancaster, preserved at Grimsthorpe, Dublin, 1907
    Calendar of the manuscripts of the Marquis of Salisbury, preserved at Hatfield House, 22 vols., London, 1883–1973
Jackson, C. (ed.). 'Yorkshire diaries', *Publications of the Surtees Society*, 65 (1875)
La Rochefoucauld, F. de. *A Frenchman in England 1784, being the Mélanges sur l'Angleterre of François de la Rochefoucauld*, ed. J. Marchant, trans. S.C. Roberts, Cambridge, 1933
Lodge, A.C. (ed.). *The account book of a Kentish estate, 1616–1704*, London, British Academy, 1927
Lumb, G.D. (ed.). 'Extracts from the *Leeds Intelligencer* and the *Leeds Mercury*, 1769–1776', *Publications of the Thoresby Society*, 33 (1938)
Melling, E. (ed.). *Kentish sources 3: aspects of agriculture and industry*, Maidstone, Kent County Council, 1961
Penny, N. (ed.). *The household account book of Sarah Fell of Swarthmoor Hall, Lancashire*, Cambridge, 1920
Prior, C.M. *Early records of the thoroughbred horse*, London, 1924
    The royal studs of the 16th and 17th centuries together with a reproduction of the 2nd Earl of Godolphin's stud book, London, 1935

Robinson, C.B. 'Rural economy in Yorkshire in 1641. Being the farming and account
    books of Henry Best of Elmswell in the East Riding', *Publications of the Surtees
    Society*, 33 (1857)
Scott-Robinson, Canon. 'The expense book of James Master Esq., 1646–1676', *Arch.
    Cantania*, 15 (1883), 152–216, 16 (1886), 241–59, 17 (1887), 321–52, 18
    (1889), 114–56
Steer, F.W. *Farm and cottage inventories of mid-Essex, 1635–1749*, Essex Record
    Office, 1950
Tawney, R.H. and Power, E. (eds.). *Tudor economic documents*, 3 vols., London, 1924
Thirsk, J. and Cooper, J.P. (eds.). *Seventeenth-century economic documents*, Oxford,
    1972
Tomkins, G.W. (ed.). 'The Tomkins diary', *Sussex Arch. Collections*, 71 (1930)
Turner, E. (ed.). 'The farming diary of Richard Stapley of Twineham, Sussex, 1682–
    1724', *Sussex Arch. Collections*, 2 (1849), 102–28
Whiting, C.E. (ed.). 'Two Yorkshire diaries: the diary of Thomas Jessop and Ralph
    Ward's journal', *Yorks. Arch. Soc. Record Series*, 117 (1952)
Zell, M. 'Accounts of a sheep and corn farm, 1558–1560', *AHR*, 27 (1979), 122–8

Government papers and publications

Acts of the Privy Council, vols. 6, 10, 12, 33
Calendar of letters and papers, foreign and domestic, Henry VIII, vols. 1, 2, 15, 21
Calendar of state papers, domestic, Edward VI, Mary, Elizabeth, 1547–80
Firth, C.H. and Rait, R.S. (eds.). *Acts and ordinances of the interregnum, 1642–1660*, 3
    vols., London, 1911
House of Commons parliamentary report, *Report from the committee upon the petitions
    relating to the false winding of wool, and the marking of sheep with pitch and tar*,
    London, 1752
    *Report of the committee to whom the petition of several gentlemen, farmers and other
    persons, breeders and feeders of sheep in the county of Lincoln, was referred*, London,
    1755
Hughes, P.L. and Larkin, J.F. (eds.). *Tudor royal proclamations*, 2 vols., Yale, 1964 and
    1969
Notestein, F.W., Relf, F.H. and Simpson, H. *Commons debates 1621*, New Haven,
    1935
Pickering, D. *The statutes at large from Magna Carta to the end of the eleventh parliament
    of Great Britain anno 1761*, 20 vols., Cambridge, 1762
Statutes of the realm, 22 Henry VIII c. 7; 23 Henry VIII c. 16; 27 Henry VIII c. 6; 31
    Henry VIII c. 7; 32 Henry VIII c. 13; 33 Henry VIII c. 5 and c. 17; 1 Edward VI c.
    5; 1 Elizabeth c. 18; 8 Elizabeth c. 8; 15 Charles II c. 7; 18 Charles II c. 2; 20
    Charles II c. 7; 22 Charles II c. 13; 32 Charles II c. 2

Primary printed sources

Anderson, J. *Essays relating to agriculture and rural affairs by a farmer*, Edinburgh,
    1775
Anon. 'An account of the extraordinary size of neat cattle and sheep in Teesvale and
    the neighbouring Yarm', *Museum Rusticum*, 6: 47 (1766)
    'Housing of great cattle', *Museum Rusticum*, 5: 25 (1765)
    'Obituary of Robert Bakewell', *Gentleman's Magazine*, 65 (1795), 969–71

'On the breeding of horses', *De re rustica*, 2 (1770), 90–1

'On the pernicious practice of breeding from blind stallions', *Museum Rusticum*, 3: 41 (1765)

The Scots farmer or selected essays on agriculture adapted to the soil and climate of Scotland, 2 vols., Edinburgh, 1773–4

Bailey, J. and Culley, G. *General view of the agriculture of the county of Cumberland*, London, 1794

General view of the agriculture of the county of Northumberland, London, 1794

Bakewell, R. *Observations on the influence of soil and climate upon wool . . .*, London, 1808

Ball, J. *The farmer's complete guide through all the articles of his profession*, London, 1760

Banks, Sir J. 'Account of twelve Lincoln sheep', *Annals of Agriculture*, 15 (1791), 357–61

Beale, J. 'Some inquiries and suggestions concerning salt for domestic uses, and concerning sheep, to preserve and improve the race', *Phil. Trans. Royal Society*, 9 (1674), 48–53

Bedford, Duke of, 'An experiment comparing four breeds of sheep', *Annals of Agriculture*, 23 (1795), 456–67

'Experiment on the comparison of four breeds of sheep', *Annals of Agriculture*, 26 (1796), 412–37

Berenger, R. *The history and art of horsemanship*, 2 vols., London, 1771

Berry, Rev. H. *Improved shorthorns and their pretentions stated, being an account of this celebrated breed of cattle*, 2nd edn, London, 1830

Billingsley, J. *General view of the agriculture of the county of Somerset*, London, 1794

Bishton, J. *General view of the agriculture of the county of Shropshire*, London, 1794

Blundeville, T. *The fower chiefyst offices belonging to horsemanshippe*, London, 1565

Boys, J. *General view of the agriculture of the county of Kent*, London, 1794

Boys, J. and Ellman, J. 'Agricultural minutes taken during a ride through Rutland and Leicester', *Annals of Agriculture*, 19 (1793), 72–146

Bradley, R. *A complete body of husbandry*, London, 1727

The gentleman annd farmer's guide for the increase and improvement of cattle, London, 1729

Brown, T. *General view of the agriculture of the county of Derbyshire*, London, 1794

Browne, Sir T. *Pseudoxia epidemica (Vulgar errors)*, London, 1646, in Keynes, G. (ed.), *Works of Sir Thomas Browne*, 4 vols., London, 1964, vol. 4

Buffon, G.L. le Clerc. *The natural history of the horse . . . with full directions for breeding and improving these useful creatures, translated from the french*, London, 1762

Campbell, J.H. 'Answers to queries relating to the agriculture of Lancashire', *Annals of Agriculture*, 20 (1793), 109–53

Carew, R. *Survey of Cornwall*, 1602, Adams and Dart reprint, London, 1969

Carter, W. *A summary of certain papers about wooll . . .*, London, 1685

Cavendish, W., Duke of Newcastle. *A new method and extraordinary invention to dress horses . . .*, London, 1667

Claridge, J. *General view of the agriculture of the county of Dorset*, London, 1793

Clifford, C. *The schoole of horsemanship*, London, 1585

Country gentleman. *The complete grazier: or gentleman and farmer's directory*, 4th edn, London, 1776

Culley, G. 'Description of the breed of sheep in the possession of Messrs Culley of Northumberland . . .', appendix 6 to Redhead, W. et al., *Observations on the different breeds of sheep*, Edinburgh, 1792

Observations on livestock, 4th edn, London, 1807

'An account of the pedigree of Mr Culley's tups', *Annals of agriculture*, 28 (1797)

Davis, R. *General view of the agriculture of the county of Oxfordshire*, London, 1794

Davis, T. *General view of the agriculture of the county of Wiltshire*, London, 1794

Defoe, D. *A plan of the English commerce*, London, 1728

   A tour through the whole island of Great Britain, 1724–1726, London, 1971

Dickson, R.W. and Stevenson, W. *General view of the agriculture of the county of Lancashire*, London, 1815

Digby, Sir K. *Two treatises, in one of which the nature of bodies, in the other the nature of man's soule, is looked into, in way of discovery of the immortality of reasonable souls*, Paris, 1644

Donaldson, J. *General view of the agriculture of the county of Northampton*, Edinburgh, 1794

Dossie, R. *Memoirs of agriculture and other oeconomical arts*, 3 vols., London, 1768–82

Driver, A. and Driver, W. *General view of the agriculture of the county of Hampshire*, London, 1794

Ellis, W. *A compleat system of experienced improvements made on sheep, grass-lambs and house-lambs*, London, 1749

   The modern husbandman or practice of farming, 8 vols., London, 1750

   The practical farmer or the Hertfordshire husbandman, London, 1738

Fitzherbert, Master. *The boke of husbandry*, 1533, ed. W.W. Skeat, English Dialect Society, London, 1882

Fraser, R. *General view of the agriculture of the county of Devon*, London, 1794

Grainger, J. *General view of the agriculture of the county of Durham*, London, 1794

Hale, T. *A compleat body of husbandry . . .*, London, 1756

Harrison, W. 'Description of England', appended to *Holinshed's chronicles* of 1586, 6 vols., London, 1807

Hartlib, S. *Samuel Hartlib his legacie or an enlargement of the discourse of husbandry used in Brabant and Flanders*, London, 1651

Harvey, W. *Anatomical exercitations concerning the generation of living creatures*, trans. M. Llewellyn, London, 1653

Highmore, N. *History of generation . . .*, London, 1651

Home, H. (Lord Kames). *The gentleman farmer: being an attempt to improve agriculture by subjecting it to rational principles*, London, 1776

Hope, W. *The perfect mareschal or compleat farrier . . . of sieur de Solleysel*, Edinburgh, 1746

Houghton, J. (ed.). *A collection of letters for the improvement of husbandry and trade*, London, 1681–4

   A collection for the improvement of husbandry and trade, London, 1692–1703

Howlett, Rev. D. 'Answers to queries', *Annals of Agriculture*, 15 (1791), 260–5

J.H. 'A report on a visit to Mr Bakewell's farm', *Gentleman's Magazine*, 63 (1793), 792–5

Jacob, G. *The country gentleman's vade mecum . . .*, London, 1717

Kent, N. *General view of the agriculture of the county of Norfolk*, London, 1794

Ker, A. *Report to Sir James Sinclair . . . of the state of sheep farming along the eastern coast of Scotland, and the interior parts of the Highlands*, Edinburgh, 1791

Keynes, G. (ed.). *Works of Sir Thomas Browne* (see Browne)

Knight, T.A. 'An account of Herefordshire breeds of sheep, cattle, horses and hogs', *Communications to the Board of Agriculture*, vol. 2 (1800), 172–91

Laurence, J. *A new system of agriculture*, London, 1726

Leatham, I. *General view of the agriculture of the East Riding of Yorkshire*, London, 1794

Lisle, E. *Observations in husbandry, being a published record of his observations on general agriculture during the period 1710–1722*, London, 1757

Low, D. *The breeds of the domestic animals of the British Isles*, London, 1842

Lowe, R. *General view of the agriculture of the county of Nottingham*, London, 1794

Luccock, J. *An essay on wool, containing a particular account of the English fleece*, London, 1809

Markham, G. *Cavalarice or the English horseman*, London, 1625
'The English housewife' (1637), in *A way to get wealth*, London, 1638
How to chuse, ride, train, and diet both hunting and running horses . . ., London, 1599
'Cheap and good husbandry' (1631), in *A way to get wealth*, London, 1638

Marshall, W. *The rural economy of Gloucestershire*, 2 vols., Gloucester, 1789
The rural economy of the Midland counties, 2 vols., London, 1790
The rural economy of Norfolk, 2 vols., London, 1787
The rural economy of the southern counties, 2 vols., London, 1798
The rural economy of the West of England, 2 vols., London, 1796
The rural economy of Yorkshire, 2 vols., London, 1788

Maxwell, G.C. *Observations on . . . growing wool in Scotland*, Edinburgh, 1756

Merret, C. 'An account of several observables in Lincolnshire, not taken notice of in Camden, or any other author', *Phil. Trans. Royal Society*, 19 (1695–7), 343–53

Monk, J. *General view of the agriculture of the county of Leicestershire*, London, 1794

Morgan, N. *The horseman's honour or the beautie of horsemanship . . .*, London, 1620

Mortimer, J. *The whole art of husbandry . . .*, London, 1707

Mouffet, T. *Health's improvement . . . or rules . . . for preparing all sorts of food*, ed. C. Bennet, London, 1655

Munn, J. *Observations on British wool . . .*, London, 1738

Osmer, W. *A treatise on the diseases and lameness of horses . . .*, London, 1761

Parkinson, R. *The experienced farmer . . .*, 2 vols., London, 1798
General view of the agriculture of the county of Rutland, London, 1808

Pennant, T. *British zoology*, 4 vols., London, 1768–70

Peters, M. (ed.). *De re rustica; or the repository for select papers on agriculture, arts and manufactures*, 2 vols., London, 1769–70

Philip, 'On the choice of horses and mares for breeding', in anon., *The Scots farmer*, 1 (1773), 321–2

Pitt, W. *General view of the agriculture of the county of Leicestershire*, London, 1809
General view of the agriculture of the county of Staffordshire, London, 1794
'Reply to queries', *Annals of Agriculture*, 15 (1791), 317–21

Plot, R. *The natural history of Staffordshire . . .*, Oxford, 1686

Pringle, A. *General view of the agriculture of the county of Westmorland*, London, 1794

Proctor-Alderdon, R. 'Devonshire sheep', *Annals of Agriculture*, 17 (1792), 299–300

Ray, J. (ed.). *Ornithology of Francis Willoughby*, London, 1678
The wisdom of God manifested in the works of creation, London, 1691

Redhead, W., Laing, R. and Marshall, W. *Observations respecting the different breeds of sheep in some of the principal counties of England*, Edinburgh, 1792

Rennie, Brown and Shirreff. *General view of the agriculture of the West Riding of Yorkshire*, London, 1794

Ross, A. *Arcana microcosmi or the hid secrets of man's body discovered*, London, 1651

Ruricolo Glocestris. 'A letter to the editors on the necessity of breeding colts from sound stallions', *Museum Rusticum*, 2: 79 (1764)

Smith J. *England's improvement revived, digested into six books*, London, 1670

Smith, J. *Chronicum rusticum commerciale or memoirs of wool . . .*, 2 vols., London, 1747
A review of the manufacturer's complaints against the wool growers, by the author of memoirs of wool, London, 1753
Snape, A. *Anatomy of an horse*, London, 1683
Society for the Encouragement of Arts, Manufactures and Commerce. *Museum rusticum et commerciale . . .*, 6 vols., 1764–6
Stone, T. *General view of the agriculture of the county of Lincoln*, London, 1794
Strickland, W. 'Reply to queries', *Annals of Agriculture*, 15 (1791), 310–17
Thomas, W. *General view of the agriculture of the county of Worcestershire*, London, 1794
Topsell, E. *The historie of four-footed beasts and serpents . . .*, London, 1607
Tuke, J. *General view of the agriculture of the North Riding of Yorkshire*, London, 1794
Turbeville, G. *The noble arte of venerie or hunting* (1576), Tudor and Stuart Library, Oxford, 1908
Turner, G. *General view of the agriculture of the county of Gloucestershire*, London, 1794
Tusser, T. *Five hundred points of good husbandry*, London, 1610
Vancouver, C. *General view of the agriculture of the county of Cambridge*, London, 1794
General view of the agriculture of the county of Essex, London, 1795
Wall, R. *A dissertation on the breeding of horses upon philosophical and experimental principles . . .*, London, 1758
Warner, Rev. Mr. *General view of the agriculture of the Isle of Wight*, London, 1794
Wedge, J. *General view of the agriculture of the county of Warwickshire*, London, 1794
General view of the agriculture of the county of Cheshire, London, 1794
White, G. *The natural history of Selborne*, (1789), Oxford, 1974
Young, A. 'Experiments in weighing fatting cattle alive', *Annals of Agriculture*, 14 (1790), 140–63
'Experiments on the introduction of the Southdown sheep into Suffolk', *Annals of Agriculture*, 15 (1791), 286–310
The farmer's tour through the east of England, 4 vols., London, 1771
General view of the agriculture of the county of Lincolnshire, London, 1813
General view of the agriculture of the county of Suffolk, London, 1794
General view of the agriculture of the county of Sussex, London, 1794 and revised 2nd edn, London, 1813
'Miscellanies in Hampshire', *Annals of Agriculture*, 23 (1795), 163–79
'A month's tour to Northamptonshire, Leicestershire . . .', *Annals of Agriculture*, 16 (1791), 480–607
On the husbandry of three celebrated farmers, Messrs Bakewell, Arbuthnot and Ducket . . ., London, 1811
A six months tour through the north of England, 4 vols., London, 1770
'Some farming notes in Essex, Kent and Sussex', *Annals of Agriculture*, 20 (1793), 220–97
'A ten days tour to Mr Bakewell's', *Annals of Agriculture*, 6 (1786), 452–502
'A tour in Sussex', *Annals of Agriculture*, 11 (1789), 170–304
'A tour through Sussex, 1793', *Annals of Agriculture*, 22 (1794), 171–334 and 494–631

Secondary sources

Abercrombie, M., Hickman, C.J. and Johnson, M.L. *A dictionary of biology*, Penguin, 1951 (and subsequent editions)

Allison, K.J. 'Flock management in the 16th and 17th centuries', *EHR*, 2nd series, 11 (1958–9), 98–112
  'The wool supply and the worsted cloth industry in Norfolk in the 16th and 17th centuries, part 1: the sheep husbandry', unpublished Ph.D. thesis, University of Leeds, 1955
Armitage, P.L. 'A preliminary description of British cattle from the late 12th to the early 16th century', *Ark*, 7 (1980), 405–13
Ashmore, O. 'Inventories as a source of local history 2: farmers', *Amateur Historian*, 4 (1959), 186–95
Baker, D.A. 'Agricultural prices, production and marketing with special reference to the hop industry in north-east Kent, 1680–1760', unpublished Ph.D. thesis, University of Kent, 1976
Berry, R.J. 'The genetical implications of domestication in animals', in P.J. Ucko and G.W. Dimbleby (eds.), *The domestication and exploitation of plants and animals*, Duckworth, 1969, pp. 207–18
Bettey, J.H. 'The development of water meadows in Dorset during the 17th century', *AHR*, 25 (1977), 37–43
Bouch, C.M.L., Jones, G.P. and Brunskill, R.W. *A short economic history of the Lake counties, 1500–1830*, Manchester University Press, 1961
Bowden, P.J. *The wool trade in Tudor and Stuart England*, Macmillan, 1962
  'Wool supply and the woollen industry', *EHR*, 2nd series, 9 (1956–7), 44–58
Bowler, P.J. 'Preformation and pre-existence in the 17th century: a brief analysis', *Journal of the History of Biology*, 4 (1971), 221–44
Bridbury, A.R. *Economic growth: England in the later Middle Ages*, Harvester Press, 1975
  '16th-century farming', *EHR*, 2nd series, 27 (1974), 538–56
Brigg, M. 'The forest of Pendle in the 17th century', *Trans. Hist. Soc. Lancs. and Cheshire*, 113 (1962), 65–96
Broad, J. 'Alternate husbandry and permanent pasture in the Midlands, 1650–1800', *AHR*, 28 (1980), 77–89
Broderick, A.H. (ed.). *Animals in archaeology*, Barrie and Jenkins, 1972
Calder, A. 'The role of inbreeding in the development of the Clydesdale breed of horses', *Proc. Royal Soc. of Edinburgh*, 47 (1927), 118–40
Carlyle, W.J. 'The changing distribution of breeds of sheep in Scotland, 1795–1965', *AHR*, 27 (1979), 19–29
Clarkson, L.A. 'The organisation of the English leather industry in the late 16th and early 17th centuries', *EHR*, 2nd series, 13 (1960), 245–56
  The pre-industrial economy in England, 1500–1750, Batsford, 1971
Cole, F.J. *Early theories of sexual generation*, Clarendon Press, 1930
Cornwall, J. 'Farming in Sussex, 1560–1640', *Sussex Arch. Collections*, 92 (1954), 48–92
Darwin, C.R. *The variation of animals and plants under domestication*, John Murray, 1868
De Lacey Mann, J. *The cloth industry in the West of England, 1640–1880*, Oxford University Press, 1971
Dent, A.A. *Cleveland bay horses*, J.A. Allen, 1978
Donkin, R.A. 'Cattle on the estates of medieval Cistercian monasteries in England and Wales', *EHR*, 2nd series, 15 (1962), 31–53
Egerton, J. *British sporting and animal paintings, 1655–1867*, Paul Mellon Collection, 1978
  George Stubbs, 1724–1806, Tate Gallery Publications, 1984

Falconer, D.S. *Introduction to quantitative genetics*, Oliver and Boyd, 1980

Finch, M.E. 'The wealth of five Northants families, 1540–1640', *Northants Record Society*, 19 (1956)

Fisher, F.J. (ed.). *Essays in the economic and social history of Tudor and Stuart England*, Cambridge University Press, 1961

Flinn, M.W. *Origins of the industrial revolution*, Longmans, 1966

Fussell, G.E. 'Animal husbandry in 18th-century England, parts 1 and 2', *Agricultural History*, 11 (1937), 96–116, 189–214

    'Four centuries of Cheshire farming systems, 1500–1900', *Trans. Hist. Soc. Lancs. and Cheshire*, 106 (1954), 57–77

    'The size of English cattle in the 18th century', *Agricultural History*, 3 (1929), 160–81

Fussell, G.E. and Goodman, C. '18th-century estimates of British sheep and wool production', *Agricultural History*, 4 (1930), 131–51

Garner, F.H. *The cattle of Britain*, Longmans, Green and Co., 1945

Gasking, E. *Investigations into generation, 1651–1828*, Hutchinson, 1967

Goodall, D.M. *A history of horse breeding*, Robert Hale, 1977

Hagedoorn, A.L. *Animal breeding*, 6th edn, Crosby Lockwood, 1962

Haldane, A.R.B. *The drove roads of Scotland*, David and Charles, 1973

Hammond, Sir J. *Hammond's farm animals*, 4th edn, Edward Arnold, 1971

Hartwell, R.M. 'A revolution in the character and destiny of British wools', in N.B. Harte and K.G. Ponting (eds.), *Textile history and economic history*, Manchester University Press, 1973, pp. 320–38

Havindon, M.A. 'Agricultural progress in open-field Oxfordshire', in E.L. Jones (ed.), *Agriculture and economic growth in England, 1650–1815*, Methuen, 1967, pp. 66–79

Hewitt, H.J. 'Medieval Cheshire', *Chetham Society Publications*, new series, 88 (1929)

Houseman, W. 'Robert Bakewell', *Journal of the Royal Agric. Soc. of England*, new series, 5 (1894), 1–31

Howell, D. 'Landlords and estate management in Wales', in J. Thirsk (ed.), *Agrarian history of England and Wales, vol. 5, 1640–1750*, Cambridge University Press, 1985, pp. 252–97

Jacob, F. *The logic of living systems*, trans. B.E. Spillman, Allen Lane, 1974

Johanssen, I. and Rendel, J. *Genetics and animal breeding*, Oliver and Boyd, 1968

Jones, E.L. (ed.). *Agriculture and economic growth in England, 1650–1815*, Methuen, 1967

    'The entry of the Southdown sheep into the Wessex chalklands', *Journal of the Agric. Club of the University of Reading* (1960)

    'Hereford cattle and Ryeland sheep, economic aspects of breed changes, 1780–1870', *Trans. Woolhope Naturalists Field Club*, 38 (1965), 36–48

Kearney, H. *Science and change, 1500–1700*, Weidenfeld and Nicolson, 1971

Kerridge, E. *The agricultural revolution*, George Allen and Unwin, 1967

    'Wool growing and wool textiles in medieval and early modern times', in J.G. Jenkins (ed.), *The wool textile industry in Great Britain*, Routledge, 1972, pp. 19–33

Keynes, G. *The life of William Harvey*, Oxford University Press, 1978

Lane, C. 'The development of pastures and meadows during the 16th and 17th centuries', *AHR*, 28 (1980), 18–30

Leadam, I.S. *Select cases in the star chamber: vol. 2, 1509–1544*, Seldon Society Publications, 1911

Lerner, I.M. and Donald, H.P. *Modern developments in animal breeding*, Academic Press, 1966

Long, W.H. 'Regional farming in 17th-century Yorkshire', *AHR*, 8 (1960), 103–24

Loudon, A. and Fletcher, J. 'Monarch of the farm', *New Scientist*, 99 (14 July 1983), 88–92

Macdonald, S. 'The role of George Culley of Fenton in the development of Northumberland agriculture', *Arch. Aeliana*, 5th series, 3 (1974), 131–41

Mason, I.L. 'The role of natural and artificial selection in the origin of breeds of farm animals', *Z. Tierzuchtg. Zuchsbiologie*, 90 (1973), 229–44

Mayer, A.W. *An analysis of the 'De Generatione Animalium' of William Harvey*, Stanford University Press, 1936

Mercer, W.B. 'William Tompson: a record of georgian farming', *Min. Agric. Journal*, 45 (1939), 1125–32

Merlen, R.H. *De canibus: dog and hound in antiquity*, J.A. Allen, 1971

Munro, J.H. 'Wool price schedules and the quality of English wools in the later middle ages *c.* 1270–1499', *Textile History*, 9 (1978), 118–69

National Sheep Breeders Association. *British sheep*, London, 1968

Nichols, J.E. *Livestock improvement in relation to heredity and environment*, 4th edn, Oliver and Boyd, 1957

Olby, R.C. *Origins of Mendelism*, Schocken Books, 1967

Orwin, C.S. and Orwin, C.S. *The open fields*, 3rd edn, Clarendon Press, 1967

Outhwaite, R.B. *Inflation in Tudor and early Stuart England*, Macmillan, 1969

Palliser, D.M. 'Tawney's century 1540–1640', *EHR*, 2nd series, 35 (1982), 339–53

Pawson, H.C. *Robert Bakewell: pioneer livestock breeder*, Crosby Lockwood, 1957

Perkins, J.A. *Sheep farming in 18th- and 19th-century Lincolnshire*, The Society for Lincolnshire History of Archaeology, 1977

Postan, M.M. *Medieval economy and society*, Penguin, 1972

Prior, C.M. *The history of the racing calendar and stud book*, Horse and Hound, 1926

Ramsey, P. *Tudor economic problems*, Gollancz, 1963

Raven, C.E. *English naturalists from Neckham to Ray*, Cambridge University Press, 1947

Rogers, J.E.T. *A history of agriculture and prices in England from 1259–1793*, 7 vols., Clarendon Press, 1866–1902

Rowe, D.J. 'The Culleys, Northumberland farmers 1767–1813', *AHR*, 19 (1971), 156–74

Russell, N.C. 'Animal breeding in England *c.* 1500–1770', unpublished Ph.D. thesis, University of London, 1981

Ryder, M.L. 'British medieval sheep and their wool types', in D.W. Crossley (ed.), *Medieval industry*, Council for British Archaeology, Report no. 40 (1981)

'A History of sheep breeds in Britain', *AHR*, 12 (1964), 1–12, 65–82

'Medieval sheep and wool types', *AHR*, 32 (1984), 14–28

'The origins and history of British breeds of sheep', *Ark*, 3 (1976), 166–72

'A Survey of European primitive breeds of sheep', *Ann. Génét. Sél. Anim.*, 13 (1981), 381–418

'Wool in 1641', *Journal of the Bradford Textile Society* for 1969–70 (1971), 36–41

Scott, J.P. 'Evolution and domestication of the dog', *Evolutionary Biology*, 2 (1968)

Smith, E. 'A 200-year-old Leicestershire farm account book', *Leicestershire Historian*, 1 (1968), 76–81

Stubbe, H. *History of genetics*, trans. T.R.W. Waters, MIT Press, 1972

Terrill, C.E. '50 years progress in sheep breeding', *Journal of Animal Science*, 17 (1958), 944–59

Thirsk, J. (ed.). *The agrarian history of England and Wales*, vol. 4, Cambridge University Press, 1967

'Agrarian history, 1540–1950', in W.G. Hoskins and R.A. McKinley (eds.), *Victoria history of the county of Leicestershire*, Oxford University Press, 1954

Horses in early modern England: for service, for pleasure, for power, University of Reading, 1978

Titow, J.Z. *English rural society, 1200–1350*, Allen and Unwin, 1972

Trow-Smith, R. *A history of British livestock husbandry*, 2 vols., Routledge, 1957 and 1959

Tupling, G.H. *The economic history of Rossendale*, Manchester University Press, 1927

Vormizzer, P.J. *Charles Darwin: the years of controversy*, University of London Press, 1972

Wallace, R. and Scott-Watson, J.A. *Farm livestock of Great Britain*, Edinburgh, 1923

Watson, J.A.S. 'Bakewell's legacy', *Journal of the Royal Agric. Soc. of England*, 89 (1928), 22–32

White, K.D. *Roman farming*, Thames and Hudson, 1970

Willett, P. *An introduction to the thoroughbred*, Stanley Paul, 1975

Wright, S. 'Mendelian analysis of the pure breeds of livestock 2: the duchess family of Shorthorns as bred by Thomas Bates', *Journal of Heredity*, 14 (1923), 405–22

Wrigley, E.A. and Schofield, R.S. *The population history of England, 1541–1871*, Edward Arnold, 1981

Zeuner, F.E. *A history of domesticated animals*, Hutchinson, 1963

Zirkle, C. 'The early history of the idea of inheritance of acquired characters and pangenesis', *Trans. American Phil. Soc.*, new series, 35 (1946), 91–151

Index